WITHDRAWN

The Role of the Spectrum
in the Cyclic Behavior
of Composition Operators

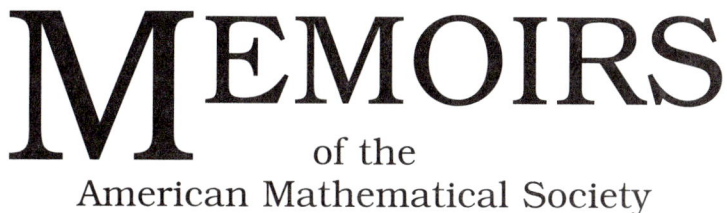
of the
American Mathematical Society

Number 791

The Role of the Spectrum
in the Cyclic Behavior
of Composition Operators

Eva A. Gallardo-Gutiérrez
Alfonso Montes-Rodríguez

January 2004 • Volume 167 • Number 791 (first of 5 numbers) • ISSN 0065-9266

American Mathematical Society
Providence, Rhode Island

2000 *Mathematics Subject Classification.* Primary 47B38, 47A16; Secondary 30D55, 30D05.

Library of Congress Cataloging-in-Publication Data

Gallardo-Gutiérrez, Eva A., 1973–
  The role of the spectrum in the cyclic behavior of composition operators / Eva A. Gallardo-Gutiérrez, Alfonso Montes-Rodríguez.
   p. cm. — (Memoirs of the American Mathematical Society, ISSN 0065-9266 ; no. 791)
  "January 2004, volume 167, number 791 (first of 5 numbers)."
  Includes bibliographical references.
  ISBN 0-8218-3432-0 (alk. paper)
  1. Function spaces. 2. Linear operators. 3. Functions of complex variables. 4. Hypergeometric functions. I. Montes-Rodríguez, Alfonso, 1961– II. Title. III. Series.

QA3.A57   no. 791
[QA323]
510 s—dc22
[515′.7246]   2003061940

Memoirs of the American Mathematical Society

This journal is devoted entirely to research in pure and applied mathematics.

Subscription information. The 2004 subscription begins with volume 167 and consists of six mailings, each containing one or more numbers. Subscription prices for 2004 are $583 list, $466 institutional member. A late charge of 10% of the subscription price will be imposed on orders received from nonmembers after January 1 of the subscription year. Subscribers outside the United States and India must pay a postage surcharge of $31; subscribers in India must pay a postage surcharge of $43. Expedited delivery to destinations in North America $35; elsewhere $130. Each number may be ordered separately; *please specify number* when ordering an individual number. For prices and titles of recently released numbers, see the New Publications sections of the *Notices of the American Mathematical Society*.

Back number information. For back issues see the *AMS Catalog of Publications*.

Subscriptions and orders should be addressed to the American Mathematical Society, P. O. Box 845904, Boston, MA 02284-5904, USA. *All orders must be accompanied by payment.* Other correspondence should be addressed to 201 Charles Street, Providence, RI 02904-2294, USA.

Copying and reprinting. Individual readers of this publication, and nonprofit libraries acting for them, are permitted to make fair use of the material, such as to copy a chapter for use in teaching or research. Permission is granted to quote brief passages from this publication in reviews, provided the customary acknowledgment of the source is given.

Republication, systematic copying, or multiple reproduction of any material in this publication is permitted only under license from the American Mathematical Society. Requests for such permission should be addressed to the Acquisitions Department, American Mathematical Society, 201 Charles Street, Providence, Rhode Island 02904-2294, USA. Requests can also be made by e-mail to reprint-permission@ams.org.

Memoirs of the American Mathematical Society is published bimonthly (each volume consisting usually of more than one number) by the American Mathematical Society at 201 Charles Street, Providence, RI 02904-2294, USA. Periodicals postage paid at Providence, RI. Postmaster: Send address changes to Memoirs, American Mathematical Society, 201 Charles Street, Providence, RI 02904-2294, USA.

© 2004 by the American Mathematical Society. All rights reserved.
This publication is indexed in *Science Citation Index*®, *SciSearch*®, *Research Alert*®, *CompuMath Citation Index*®, *Current Contents*®/*Physical, Chemical & Earth Sciences*.
Printed in the United States of America.

∞ The paper used in this book is acid-free and falls within the guidelines established to ensure permanence and durability.
Visit the AMS home page at http://www.ams.org/

10 9 8 7 6 5 4 3 2 1   09 08 07 06 05 04

CONTENTS

Preface	ix
Chapter 1. Introduction and Preliminaries	1
Weighted Hardy spaces	1
The Dirichlet space	2
The Hardy space	2
Functional Hilbert spaces	2
Composition operators	3
Weighted Dirichlet spaces	3
Equivalent norms on weighted Dirichlet spaces	3
Linear fractional maps	4
Iterates	5
The Comparison Principle	6
Main results	7
Table I	8
Table II	10
Acknowledgment	12
Chapter 2. Linear Fractional Maps with an Interior Fixed Point	13
Non-elliptic linear fractional maps with an exterior fixed point	13
Elliptic automorphism	17
Interior and boundary fixed point	19
Hypercyclicity	23
The hyperbolic automorphism	24
The Hypercyclic Criterion	24
A Beurling type theorem	25
The Inner Product Criterion	25
Invariant subspaces of finite codimension	31
Chapter 3. Non Elliptic Automorphisms	35
Hypercyclicity	34
The parabolic automorphism	36
The hyperbolic automorphism	38
The failure of the spectrum	40
The eigenfunctions of C_φ with φ a parabolic automorphism	42
Cyclicity	43
The Dirichlet space of the upper half plane	44
The Hardy space of the upper half plane	44
A Paley-Wiener Theorem for $\mathcal{D}_\pi$	44

The parabolic automorphism	45
The hyperbolic automorphism	46
Chapter 4. The Parabolic Non Automorphism	**48**
Cyclicity	48
Zeros of weighted Dirichlet functions	51
Non hypercyclicity	52
An appetizer	52
The spectrum of C_φ	53
The eigenfunctions form a spanning set	53
Renorming $\mathcal{S}_\nu$	54
Isometric functional Hilbert spaces	55
A complete orthogonal system of $L^2(\mathbb{R}^+, dt/t)$	56
The extension of the isometries	59
The Laguerre polynomials	60
Chapter 5. Supercyclic Linear Fractional Composition Operators	**66**
The easy part	66
The parabolic non-automorphism	68
The Angle Criterion	68
The adjoint of C_φ	69
A three term recurrence formula	71
An integration formula	71
Endnotes	**77**
Bibliography	**79**

Abstract

A bounded operator T acting on a Hilbert space $\mathcal{H}$ is called cyclic if there is a vector x such that the linear span of the orbit $\{T^n x : n \geq 0\}$ is dense in $\mathcal{H}$. If the scalar multiples of the orbit are dense, then T is called supercyclic. Finally, if the orbit itself is dense, then T is called hypercyclic. We completely characterize the cyclicity, the supercyclicity and the hypercyclicity of scalar multiples of composition operators, whose symbols are linear fractional maps, acting on weighted Dirichlet spaces. Particular instances of these spaces are the Bergman space, the Hardy space, and the Dirichlet space. Thus, we complete earlier work on cyclicity of linear fractional composition operators on these spaces. In this way, we find exactly the spaces in which these composition operators fail to be cyclic, supercyclic or hypercyclic. Consequently, we answer some open questions posed by Zorboska. In almost all the cases, the cut-off of cyclicity, supercyclicity or hypercyclicity of scalar multiples is determined by the spectrum. We will find that the Dirichlet space plays a critical role in the cut-off.

Received by the editor February 22, 2000 and in revised form February 15, 2001.

2000 *Mathematics Subject Classification*. Primary 47B38, 47A16; Secondary 30D55, 30D05.

Key words and phrases. Composition operator, cyclic operator, supercyclic operator, hypercyclic operator, spectrum, Fourier transform, Haar measure, zeros of holomorphic functions, Laguerre polynomials.

This work was supported partially by Ministerio de Ciencia y Tecnología, Plan Nacional I+D Ref. BFM 2000-0360 and Junta de Andalucía Ref. FQM-260. The first named author was also supported by Plan Propio de la Universidad de Cádiz.

Preface

The study of composition operators on spaces of analytic functions merges successfully with other branches of mathematics, such as Classical Function Theory or Harmonic Analysis. Considered on classical function spaces, composition operators provide new meanings to old theorems. Related to Operator Theory, the study of composition operators in the context of function spaces is part of *Concrete Operator Theory* and, therefore, they are a good source of examples to describe and generalize results to other classes of operators. In particular, composition operators are an excellent test class because they are a crosscutting class since, just as weighted shift operators, they do not fall into a class of operators satisfying a simple prescribed property.

The characterization of the different cyclic properties of scalar multiples of linear fractional composition operators on weighted Dirichlet spaces that we carry out in this work requires a large variety of techniques. They go from studying the growth of the zero sequences of holomorphic functions to the use, in a special way, of the Laguerre polynomials in functional Hilbert spaces, passing through Fourier analysis and Haar measures.

The study of the cyclic properties of an operator plays a prominent role in determining its invariant subspaces, which is what makes the concept of cyclicity so important in Operator Theory. An important instance of this fact is the Spectral Theorem for normal operators (see Radjavi and Rosenthal's book [RR]). Moreover, the connection between cyclic operators and the *Invariant Subspace Problem*, unsolved in the Hilbert space setting, has attracted the attention of many analysts in the last few decades. One could think that studying the cyclicity of composition operators just induced by linear fractional transformations could be seen as a narrow goal. However, Nordgren, Radjavi and Rosenthal [NRR] have shown the equivalence of the general Invariant Subspace Problem to a problem about the invariant subspaces of a composition operator induced by a hyperbolic automorphism of the unit disk, that is, induced by a particular linear fractional transformation (see also Nordgren, Rosenthal and Wintrobe's paper [NRW]).

The knowledge of any property of linear fractional composition operators becomes important when many results on composition operators induced by analytic self-maps of the unit disk seem to be derived from it. Roughly speaking, any analytic self map of the unit disk, except the elliptic automorphisms or maps for which the Denjoy-Wolff point is a critical point, can be classified according to a linear fractional model by means of an intertwining relation (see Cowen and MacCluer's book [CM, Section 2.4]). In some sense, it is expected that the properties of the composition operator induced by the linear fractional model are transferred by means of the intertwining relation to the original map. Of course, this does not necessarily mean that this transference is an easy task. As an example of what we have just said, we point out the use of the Linear Fractional Model Theorem to study the compactness of general composition operators (see Shapiro's book [Sh1, Chap. 9]).

The Linear Fractional Model Theorem seems to be especially useful in those processes that involve iteration, as is the case with the different cyclic properties. Bourdon and Shapiro [BM2] carried out a nearly complete program of transferring the cyclic and hypercyclic properties of linear fractional composition operators to general composition operators on the Hardy space. We hope that the study of the linear fractional case should be the key to the general case of cyclicity and composition operators.

In all this work, the spectrum will play a central role. It is known to the experts that the properties invariant under similarities, as it is the case for the cyclicity, determine quite a lot about the spectral picture of an operator (see Herrero's book [He1]). A trivial example of this fact is that if the spectrum of an operator is a segment, then the operator is not compact. With respect to the cyclic properties, many things related to the spectrum are known. For instance, each component of the spectrum of a hypercyclic operator must meet the unit circle [Ki] or, a rich supply of eigenvectors provides the operator with more opportunities to be cyclic or even hypercyclic, (see the work of Clancey and Rogers [CR] and Godefroy and Shapiro [GS]). While it is true that in some instances, like the Hardy space or the Bergman space, we could have used spectral techniques to prove whether or not a given composition operator satisfies a cyclic property, in general, we have avoided proceeding in this way for several reasons. On the one hand, in many situations, especially when the underlying space is small, there is a complete lack of eigenvectors and, on the other hand, in many of the main instances that arise here, the spectrum of the linear fractional composition operator is unknown. We point out here that the limit cases are not determined by the spectrum. For instance the identity operator, which is the least hypercyclic operator, satisfies all the spectral properties of hypercyclic operators. The proofs that we present here, will provide a unified treatment of each linear fractional map in a large variety of spaces.

On the opposite direction, many of the results we obtain suggest what the spectrum of some linear fractionally induced composition operators should be. From the point of view of Concrete Operator Theory, this work shows to what extent it is true that the spectrum determines the cyclic behavior of this significant class of operators.

Besides the intuition that the spectrum of an operator provides, sometimes the spectrum (or what it should be) will determine the cyclicity, other times the spectrum will be used as a fundamental technical tool in the proofs and sometimes (cyclicity of composition operators induced by non elliptic automorphisms) the spectrum will apparently play no role at all.

Eva A. Gallardo and Alfonso Montes

CHAPTER 1

INTRODUCTION AND PRELIMINARIES

In this preliminary chapter we introduce the Hilbert spaces of holomorphic functions in which our work is set. Linear fractional composition operators are also presented. Those basic facts about linear fractional maps which are relevant to our study of these operators are described. Then a brief discussion follows about the Comparison Principle for the different cyclic properties. We close the chapter by stating our main results on cyclicity of composition operators induced by linear fractional maps.

WEIGHTED HARDY SPACES

Weighted Hardy spaces were introduced by Shields [Shi] to study weighted shift operators on Hilbert spaces. But also composition operators have been extensively studied on these spaces. In Cowen and MacCluer's book [CM], a large number of results about composition operators on these spaces and a good list of references can be found. Let $\mathbb{D}$ denote the open unit disk in the complex plane. For each sequence of positive numbers $\beta = \{\beta_n\}$, the weighted Hardy space $\mathcal{H}^2(\beta)$ is defined as the Hilbert space of functions $f(z) = \sum_{n=0}^{\infty} a_n z^n$ analytic on $\mathbb{D}$ for which the norm

$$\|f\|_\beta^2 = \sum_{n=0}^\infty |a_n|^2 \beta_n^2$$

is finite. Since the function $\sum_{n=0}^\infty (n\beta_n)^{-1} z^n$ is in $\mathcal{H}^2(\beta)$, we have $\limsup \beta_n^{-1/n} \leq 1$. Notice that the norm above is induced by the following inner product

$$\left\langle \sum_{n=0}^\infty a_n z^n, \sum_{n=0}^\infty b_n z^n \right\rangle = \sum_{n=0}^\infty a_n \bar{b}_n \beta_n^2,$$

and that the monomials $\{z^n\}$ form a complete orthogonal system. As a particular consequence of this fact, the polynomials are dense in $\mathcal{H}^2(\beta)$. Observe also that $\|z^n\|_\beta = \beta_n$ and that weighted Hardy spaces are natural spaces in the sense that norm convergence in $\mathcal{H}^2(\beta)$ implies uniform convergence on compact subsets of the unit disk.

A word about notation. Unless there is risk of ambiguity, any equivalent norm in any space will be denoted by $\|\cdot\|$.

We stress here that for particular choices of the sequence of weights $\{\beta_n\}$ the spaces $\mathcal{H}^2(\beta)$ turn out to be well-known function spaces. For instance, when $\beta_n = 1$ for every n, the space $\mathcal{H}^2(\beta)$ is the classical Hardy space $\mathcal{H}^2$; when $\beta_n = (n+1)^{-1/2}$, the Bergman space $\mathcal{A}^2$ and when $\beta_n = (n+1)^{1/2}$, the Dirichlet space $\mathcal{D}$.

The Dirichlet Space.
The Dirichlet space will be one of the primary settings in our work on composition operators. Indeed, we will see that most of the time the cut-off of cyclicity, supercyclicity and hypercyclicity occurs at this space. The Dirichlet space is the collection of functions analytic on $\mathbb{D}$ whose first derivatives have square integrable modulus over $\mathbb{D}$. For $f \in \mathcal{D}$ the norm in $\mathcal{D}$ has the integral representation

$$\|f\|^2 = |f(0)|^2 + \int_\mathbb{D} |f'(z)|^2 \, dA(z),$$

where $dA(z)$ is the Lebesgue area measure on $\mathbb{D}$ normalized to have unit mass. For a univalent function f the integral above is just the area of $f(\mathbb{D})$.

The Hardy space.
The Hardy space $\mathcal{H}^2$ is the space of analytic functions $f(z) = \sum_{n=0}^\infty a_n z^n$ such that

$$\|f\|^2 = \sum_{n=0}^\infty |a_n|^2$$

is finite. In this case, it is less obvious that there is also an integral representation of the norm in $\mathcal{H}^2$. The point here is that the radial limits

$$\lim_{r \to 1^-} f(re^{i\theta})$$

of any f in $\mathcal{H}^2$ define a function $f^\star(e^{i\theta})$ on $\partial \mathbb{D} = \{z \in \mathbb{C} : |z| = 1\}$ that is in $L^2(\partial \mathbb{D})$ (see [Ru1, Chap. 17] or [Dur, Chap. 2], for instance). Usually, the function $f^\star(e^{i\theta})$ is also denoted by $f(e^{i\theta})$. Now, it is not difficult to show that the norm of a function $f \in \mathcal{H}^2$ has also the integral representation

$$\|f\|^2 = \frac{1}{2\pi} \int_{-\pi}^\pi |f(e^{i\theta})|^2 \, d\theta.$$

The Hardy space will play a central role in many of the proofs in this work.

FUNCTIONAL HILBERT SPACES

DEFINITION 1.1. A functional Hilbert space $\mathcal{H}$ is a Hilbert space of complex-valued functions defined on a set X satisfying
  (i) Non-triviality: $\mathcal{H} \neq \{0\}$.
  (ii) For each $x \in X$ the point evaluation functional $f \to f(x)$ is bounded.

Any weighted Hardy space $\mathcal{H}^2(\beta)$ is a functional Hilbert space in which $X = \mathbb{D}$. The sequence of weights β_n allows us to consider the generating function

$$k(z) = \sum_{n=0}^\infty \frac{z^n}{\beta_n^2}$$

which is analytic on $\mathbb{D}$. For each $w \in \mathbb{D}$ the reproducing kernel is defined by $K_w(z) = k(\bar{w}z)$. It is easy to check that $\|K_w\|^2 = k(|w|^2)$ and for any $f(z) = \sum_{n=0}^\infty a_n z^n \in \mathcal{H}^2(\beta)$ we have the reproducing property

$$\langle f, K_w \rangle = \sum_{n=0}^\infty a_n w^n = f(w)$$

(see [CM, Thm. 2.10] for the details). The reproducing kernels are very important because they somehow define the spaces $\mathcal{H}^2(\beta)$. In fact, they span $\mathcal{H}^2(\beta)$ (see [CM, Prop. 1.2]). The reproducing kernel in the Hardy space is

$$\sum_{n=0}^{\infty} \bar{w}^n z^n = \frac{1}{1-\bar{w}z}.$$

It will be very useful in this work, even if we are working in the more general situation of weighted Hardy spaces $\mathcal{H}^2(\beta)$.

COMPOSITION OPERATORS

If φ is a holomorphic self-map of the unit disk, then it is a consequence of the Closed Graph Theorem that the composition operator C_φ defined by $C_\varphi f = f \circ \varphi$ is a continuous operator on $\mathcal{H}(\mathbb{D})$, the space of all holomorphic functions on $\mathbb{D}$ endowed with the topology of uniform convergence on compacta. However, if we consider $C_\varphi f$, with $f \in \mathcal{H}^2(\beta)$, then it is an open (and complicated) problem to determine which holomorphic self maps of the unit disk induce bounded composition operators on $\mathcal{H}^2(\beta)$. For instance, many composition operators are unbounded when the sequence of weights decreases too fast (see [CM, Thm. 5.2]).

Weighted Dirichlet spaces.

To avoid these pathologies we will restrict ourselves to the case in which the weights $\beta_n = (n+1)^\nu$, where ν is a real number. These spaces are also known in the literature as weighted Dirichlet spaces or $\mathcal{S}_\nu$ spaces (see [CM] or [Zo2], for instance). Observe that we still retain the Hardy, Bergman and Dirichlet spaces in the framework of our study. Having said this, sometimes we find it interesting to state our results for general weights.

We observe that the bigger ν is, the smaller the space $\mathcal{S}_\nu$ is. By "small" we mean that the functions are of slower growth when approaching to the boundary of the unit disk. Also, if $\nu_1 > \nu_2$, then $\mathcal{S}_{\nu_1}$ is strictly contained in $\mathcal{S}_{\nu_2}$. Observe that for smaller spaces than the Dirichlet space ($\nu > 1/2$) all functions in $\mathcal{S}_\nu$ extend continuously to the boundary of the unit disk.

Even on the spaces $\mathcal{S}_\nu$, not all composition operators are bounded. An obvious necessary condition for boundedness is that the inducing symbol $\varphi = C_\varphi z \in \mathcal{S}_\nu$, which is not always the case. For instance, when φ takes on each value in $\mathbb{D}$ infinitely many times, the Dirichlet norm of φ is infinite. However, the symbols that induce cyclic (and, therefore, supercyclic and hypercyclic) composition operators must be univalent. This result, for the Hardy space $\mathcal{H}^2$, was proved by Bourdon and Shapiro [BS2, p. 18], and Zorboska [Zo3] observed that the same proof works for weighted Hardy spaces. It is easy to see that the univalence of φ is a sufficient condition for the operator C_φ to be bounded on the Dirichlet space.

EQUIVALENT NORMS ON WEIGHTED DIRICHLET SPACES

In some situations it will be convenient to work with an equivalent norm on $\mathcal{S}_\nu$ that reflects the values of the functions rather than the coefficients of their Taylor developments. Recall that $dA(z) = \frac{1}{\pi} dxdy = \frac{1}{\pi} rdrd\theta$, ($z = x + iy = re^{i\theta}$) denotes the normalized Lebesgue area measure on $\mathbb{D}$. The following Lemma is well known for $\nu \leq 0$ and for $0 \leq \nu < 2$ can be found in [Zo2].

LEMMA 1.2. *Suppose that ν is a real number and $l \geq -1$ is an integer such that $\nu < l + 1$. Then the expression*

$$\|f\|^2 = \sum_{i=0}^{l} |f^{(i)}(0)|^2 + \int_{\mathbb{D}} |f^{(l+1)}(z)|^2 (1-|z|^2)^{2l+1-2\nu} dA(z) \qquad (f \in \mathcal{S}_\nu),$$

where the first term in the right hand side above does not appear when $l = -1$, defines an equivalent norm on $\mathcal{S}_\nu$.

PROOF. We set $c = 2l + 1 - 2\nu > -1$. Let $f(z) = \sum_{n=0}^{\infty} a_n z^n$ be in $\mathcal{S}_\nu$. The Taylor series representation for $f^{(l+1)}(z)$ along with the orthogonality of the monomials z^n in $L^2(\partial \mathbb{D})$ shows that

$$\int_{\mathbb{D}} |f^{(l+1)}(z)|^2 (1-|z|^2)^c dA(z) = 2 \sum_{n=l+1}^{\infty} \left[\frac{n!}{(n-l-1)!} \right]^2 |a_n|^2 \int_0^1 r^{2n-2l-1} (1-r^2)^c dr.$$

The integral on the right hand side above is well defined because $c > -1$. The change of variables $u = r^2$ shows that the above expression is equal to

$$\sum_{n=l+1}^{\infty} \left(\frac{n!}{(n-l-1)!} \right)^2 \frac{\Gamma(2l+2-2\nu)\Gamma(n-l)}{\Gamma(n+l+2-2\nu)} |a_n|^2,$$

where Γ denotes the Gamma function. Upon applying Stirling's formula we see that

$$\lim_{n \to \infty} \frac{n^2(n-1)^2 \cdots (n-l)^2 \Gamma(n-l)}{(n+1)^{2\nu} \Gamma(n+l+2-2\nu)}$$

is finite and different from zero. Finally, since the replacement of finitely many weights yields an equivalent norm, the statement of the lemma follows. □

EXAMPLE 1.3. If ψ is an analytic and bounded function on $\mathbb{D}$, then, using Lemma 1.2, it is easy to show that the multiplication operator M_ψ that assigns to each function $f \in \mathcal{S}_\nu$ the function ψf is bounded in $\mathcal{S}_\nu$ for all $\nu < 0$. This is also true for the Hardy space.

LINEAR FRACTIONAL MAPS

The simplest composition operators are those induced by linear fractional maps

$$\varphi(z) = \frac{az+b}{cz+d},$$

where a, b, c and d are complex numbers satisfying $ad - bc \neq 0$ to avoid that $\varphi(z)$ be independent of z. They extend to the extended complex plane $\mathbb{C}^\infty = \mathbb{C} \cup \{\infty\}$ by defining $\varphi(\infty) = a/c$, and $\varphi(-d/c) = \infty$ if $c \neq 0$, while $\varphi(\infty) = \infty$ if $c = 0$. In this way, they become bijections on $\mathbb{C}^\infty$. It is easy to see that a linear fractional map that fixes three different points is the identity. Linear fractional maps act transitively on triples of different points and they send circles (in $\mathbb{C}^\infty$) onto circles (in $\mathbb{C}^\infty$), see [Ah].

The linear fractional maps can be classified according to their fixed points. It is easy to see that a linear fractional map which is not the identity has one or two fixed points in the extended complex plane. Two linear fractional maps φ and ψ are said to be conjugate if there is another linear fractional map T such that

$\varphi = T^{-1}\psi T$. If φ has only one fixed point α, then it is called *parabolic*. If this is the case, we can conjugate by
$$T(z) = \frac{1}{z-\alpha}$$
and we obtain $\varphi = T^{-1}\psi T$, where $\psi(z) = z + \tau$ and $\tau \neq 0$. If φ has two distinct fixed points α and β, then $\varphi = T^{-1}\psi T$, where $\psi(z) = \mu z$ and
$$Tz = \frac{z-\alpha}{z-\beta}.$$

In this case, the linear fractional map is called elliptic if $|\mu| = 1$; hyperbolic if $\mu > 0$ and loxodromic otherwise (see [Ah] for more details). In any of the preceding cases, the map ψ is called the normal form.

A necessary condition for a linear fractional map φ to define a composition operator on a space of analytic functions on $\mathbb{D}$ is that φ takes the unit disk into itself and because linear fractional maps are univalent, this condition also insures that C_φ is bounded on any of the $\mathcal{S}_\nu$ spaces for $\nu \leq 1/2$.

In general, for $\nu \leq 0$, composition operators are bounded (see [CM, Chap. 3]). If $\nu > 0$, we have the following proposition due to Hurst ([Hu, Theorem 9]) that guarantees that linear fractional maps induce bounded composition operators on $\mathcal{S}_\nu$ for all ν.

PROPOSITION 1.3. *Let φ be a linear fractional map with $\varphi(\mathbb{D}) \subset \mathbb{D}$. Then C_φ is bounded on all the $\mathcal{S}_\nu$ spaces.*

PROOF. If $\nu \leq 0$, we already know that C_φ is bounded on $\mathcal{S}_\nu$. Since φ takes the unit disk into itself, φ' and φ'' are bounded on $\mathbb{D}$. Thus the multiplication operators $M_{\varphi'}$ and $M_{\varphi''}$ are bounded on $\mathcal{S}_\nu$ for $\nu \leq 0$. Now, suppose that we have already proved that C_φ, $M_{\varphi'}$ and $M_{\varphi''}$ are also bounded for $\mathcal{S}_\nu$ with $\nu \leq m$. Let $m < \nu \leq m+1$ and let f be in $\mathcal{S}_\nu$. It is easy to see that $f \in \mathcal{S}_\nu$ if and only if f' is in $\mathcal{S}_{\nu-1}$. Therefore, both f and f' are in $\mathcal{S}_{\nu-1}$. It follows that
$$(f\varphi')' = f'\varphi' + f\varphi''$$
is in $\mathcal{S}_{\nu-1}$. Also, since $M_{\varphi'}$ and C_φ are bounded on $\mathcal{S}_{\nu-1}$,
$$(f \circ \varphi)' = (f' \circ \varphi)\varphi'$$
is in $\mathcal{S}_{\nu-1}$. Hence, the desired result follows from the Closed Graph Theorem. □

ITERATES

Obviously, the behavior of the iterates of an operator plays a fundamental role in the study of any kind of cyclic property. Let φ_0 denote the identity map. Then $\varphi_n = \varphi \circ \varphi_{n-1}$ is the n-th iterate of φ. The equation $C_\varphi^n = C_{\varphi_n}$ relates the iterates of C_φ to the iterates of φ. Therefore, the behavior of $\{\varphi_n\}$ will be important in determining the cyclic behavior of C_φ.

The normal forms are specially useful for computing the iterates φ_n. If φ is parabolic, then $\varphi = T^{-1}\psi T$, where $\psi(z) = z + \tau$ and $\tau \neq 0$. Thus $\varphi_n = T^{-1}\psi_n T$. Since $\psi_n(z) = z + n\tau$ tends uniformly on compact subsets of $\mathbb{C}$ to ∞, we see that φ_n converges on compact subsets of $\mathbb{C}^\infty$, except the fixed point, to its fixed point.

If φ is loxodromic or hyperbolic, then one also easily checks that $\{\varphi_n\}$ converges uniformly on compact subsets of $\mathbb{C}^\infty$, minus one fixed point, to the other fixed point. The latter is called the attractive fixed point and the former the repulsive fixed point.

The chain rule proves that the value of the derivative at the fixed point is preserved under conjugation. Therefore, the derivative of a parabolic linear fractional map at its fixed point is 1; while the derivative of a hyperbolic one is strictly less than 1 at its attractive fixed point and strictly greater than 1 at its repulsive point. In the course of this work, we will see that the cyclic properties of C_φ strongly depend on how fast this convergence is.

The cyclic properties C_φ reflect the dynamics of φ. The dynamics of φ depends strongly on the derivative at its Denjoy-Wolff point: the fixed point if φ is parabolic and the attractive fixed point if φ is hyperbolic or loxodromic. The derivative is a measure of the speed of how $\{\varphi_n\}$ approaches to its Denjoy-Wolff point. This causes that the study of the cyclicity of C_φ, when φ is a parabolic non automorphism, requires a much subtler analysis.

Finally, the condition $\varphi(\mathbb{D}) \subset \mathbb{D}$ imposes some restrictions on the location of the fixed points of φ. By using the normal forms, it is easy to prove the following proposition (see [Sh2, p. 5]).

PROPOSITION 1.4. *Suppose that φ is a linear fractional map with $\varphi(\mathbb{D}) \subset \mathbb{D}$. Then*
  (a) *If φ is parabolic, then its fixed point is on $\partial \mathbb{D}$.*
  (b) *If φ is hyperbolic, the attractive fixed point is in $\overline{\mathbb{D}}$ and the other fixed point outside of $\mathbb{D}$ and both fixed points are on $\partial \mathbb{D}$ if and only if φ is an automorphism of $\mathbb{D}$.*
  (c) *If φ is loxodromic or elliptic, one fixed point is in $\mathbb{D}$ and the other fixed point outside of $\overline{\mathbb{D}}$. The elliptic ones are always automorphisms of $\mathbb{D}$. The fixed point in $\mathbb{D}$ for the loxodromic ones is attractive.*

THE COMPARISON PRINCIPLE

A bounded linear operator T acting on a linear metric space $\mathcal{E}$ is said to be *cyclic* if there is a vector $x \in \mathcal{E}$ such that

$$\overline{\text{span}} \{T^n x : n = 0, 1, \ldots\} = \mathcal{E}.$$

The vector x is called a cyclic vector for T. If there is a vector x such that the orbit

$$\{T^n x : n = 0, 1 \ldots\}$$

is dense, then T is said to be *hypercyclic* and the vector x a hypercyclic vector for T. Finally, between the concept of cyclicity and hypercyclicity there is the concept of supercyclicity. An operator T is said to be supercyclic, if there is $x \in \mathcal{E}$ such that the projective orbit

$$\{\lambda T^n x : n = 0, 1, 2 \ldots \text{ and } \lambda \in \mathbb{C}\}$$

is dense in $\mathcal{E}$. In this case the vector x is called supercyclic for T. Trivially, any operator with a hypercyclic scalar multiple is supercyclic. However, certainly not all supercyclic operators arise in this way. We will come back to this later.

A basic tool in the study of any of the cyclic properties is the following [Sh2, p. 111]. See also the work by Salas [Sa], where it was noticed that the Comparison Principle is also applicable for cyclic and supercyclic operators.

PROPOSITION 1.5 (Comparison Principle). *Suppose that $\mathcal{E}$ is a linear metric space and $\mathcal{F}$ is a dense subspace that is itself a linear metric space with a stronger topology. Suppose that T is a continuous linear transformation on $\mathcal{E}$ that also maps the smaller space $\mathcal{F}$ into itself, and is continuous in the topology of this space. If T is cyclic on $\mathcal{F}$, then it is also cyclic on $\mathcal{E}$ and has an $\mathcal{E}$-cyclic vector that belongs to $\mathcal{F}$. Furthermore, the same is true for supercyclic and hypercyclic operators.*

The proof, which is very easy, follows from the density of $\mathcal{F}$ in $\mathcal{E}$ and the fact that convergence in $\mathcal{F}$ implies convergence of $\mathcal{E}$. Since the polynomials are dense in any weighted Hardy space, the Comparison Principle applies to the case in which $\mathcal{H}^2(\beta_1)$ is contained in $\mathcal{H}^2(\beta_2)$ and convergence in $\mathcal{H}^2(\beta_1)$ implies that of $\mathcal{H}^2(\beta_2)$.

Seidel and Walsh [SW] proved that if φ is an automorphism of the unit disk without fixed point in $\mathbb{D}$, then the composition operator C_φ is hypercyclic on $\mathcal{H}(\mathbb{D})$. Since norm convergence in the Hardy space implies uniform convergence on compact subsets, the Seidel and Walsh Theorem can be regarded as a special case of a result of Bourdon and Shapiro [BS2, Thm. 2.2] that under the same hypotheses C_φ is hypercyclic on the Hardy space $\mathcal{H}^2$. The point is that the smaller the space is the better behaved the hypercyclic vectors will be, if there are any.

Another relevant example is related to Birkhoff's Theorem [Bi] that the operator of translation by a non-zero complex number is hypercyclic on the space of all entire functions endowed with the topology of uniform convergence on compacta. Chan and Shapiro [CS] showed that the translation operator is hypercyclic on Hilbert spaces of entire functions that can have arbitrarily slow growth. By the Comparison Principle, this provides better behaved hypercyclic functions in Birkhoff's Theorem. From this point of view, we will solve the corresponding problem for the Seidel and Walsh Theorem. In strong contrast with the situation in Birkhoff's Theorem, we will see that when the spaces get smaller the operators get "less cyclic". For instance, the non-elliptic automorphisms of $\mathbb{D}$ induce hypercyclic composition operators on the Hardy space but they fail to be even cyclic (in a very strong sense) on the Dirichlet space.

MAIN RESULTS

In this section we state our main results on cyclic composition operators induced by linear fractional maps. We also record what is known about this subject.

Nina Zorboska, in her dissertation [Zo1], studied composition operators induced by non-elliptic disk automorphisms, and proved that they are all cyclic on the Hardy space.

Bourdon and Shapiro (see [BS1] and [BS2]) obtained a complete characterization in the Hardy space of the cyclic and hypercyclic composition operators induced by linear fractional maps. In [BS2], using the Linear Fractional Model Theorem, Bourdon and Shapiro also made an extensive study of the cyclic behavior for composition operators (not necessarily with linear fractional symbols) on the Hardy space.

Zorboska [Zo3] has obtained some results on cyclicity and hypercyclicity on smooth weighted Hardy spaces, that is, weighted spaces whose functions have continuous first derivatives on the boundary of the unit disk (essentially weighted Hardy spaces strictly smaller than $\mathcal{S}_{3/2}$). She left open the question of what happens for bigger spaces. In particular she asks for the cut-off of cyclicity of composition operators induced by parabolic automorphisms or non-automorphisms, and hyperbolic automorphisms. In answering these questions we develop some methods that allow us to extend and improve Zorboska's results in several directions. In this work we completely characterize the cyclic and hypercyclic behavior of linear fractional composition operators, as well as their scalar multiples, on weighted Dirichlet spaces.

With respect to supercyclic operators, Ansari and Bourdon [AB] proved that composition operators whose inducing symbol has a fixed point in $\mathbb{D}$ are not supercyclic on the Hardy space $\mathcal{H}^2$. However, the nontrivial situation is that in which the inducing symbol is a parabolic non-automorphism. In this case, Shapiro [Sh3] proved that for any λ the operator λC_φ is not hypercyclic on $\mathcal{H}^2$. Finally, the authors [GM1] proved that C_φ is not supercyclic on the Hardy space. Shapiro ([Sh4]) has generalized this result to the Hardy spaces $\mathcal{H}^p$.

All the results about the cyclic behavior of composition operators on $\mathcal{S}_\nu$ spaces are summarized in Table I below.

Type of φ	Cyclic	Supercyclic	Hypercyclic	Example
Hyperbolic Automorphism	$\nu < 1/2$	$\nu < 1/2$	$\nu < 1/2$	$\dfrac{3z+1}{z+3}$
Parabolic Automorphism	$\nu < 1/2$	$\nu < 1/2$	$\nu < 1/2$	$\dfrac{(1+i)z-1}{z+i-1}$
Hyperbolic Non-Automorphism	Always	$\nu \leq 1/2$	$\nu < 1/2$	$\dfrac{1+z}{2}$
Parabolic Non-Automorphism	$\nu \leq 3/2$	Never	Never	$\dfrac{1}{2-z}$
Interior & Exterior	Always	Never	Never	$\dfrac{-z}{2+z}$
Interior & Boundary	Never	Never	Never	$\dfrac{z}{2-z}$
Elliptic Irrational Rotation	Always	Never	Never	$e^{2i/3}z$
Elliptic Rational Rotation	Never	Never	Never	$e^{2\pi i/3}z$

TABLE I

Our main contribution is to find where the exact cut-off for each kind of cyclicity and to discover the cyclic behavior on the spaces $\mathcal{S}_\nu$ when ν is between 0 and $1/2$ or between 0 and $3/2$, depending on the type of φ. In the course of this work we will see that the spectrum is what causes the cut-off of cyclicity. There is only one exception in which the cut-off is not determined by the spectrum: the non-elliptic automorphism case.

Except for parabolic non-automorphisms the cut-off, if it exists, happens at the Dirichlet space. For instance, the following theorem, that we extract from Table I, says that with respect to the cyclic behavior of composition operators there is no difference between the Hardy space and the Bergman space.

THEOREM 1.6. *Suppose that $\nu < 1/2$. A linear fractional self map of $\mathbb{D}$ induces a cyclic (respectively, supercyclic or hypercyclic) composition operator on $\mathcal{S}_\nu$ if and only if it does so on the Hardy space $\mathcal{H}^2$.*

The parabolic non-automorphism case is the only one in which the cut-off of cyclicity occurs at $3/2$ and the only one in which we have non-strict inequality there. The point is that the cut-off is very mild. In fact, C_φ is still cyclic on the spaces $\mathcal{S}_\nu$ modulo constants for $\nu > 3/2$. In any other case, the operators fail to be cyclic in a very strong sense. We say that an operator T is strongly non cyclic if the closure of the span of the orbit of any vector has infinite codimension. As a byproduct of all our results in the following Chapters we have the following

THEOREM 1.7. *If a linear fractional composition operator whose symbol is not a parabolic non-automorphism is not cyclic on $\mathcal{S}_\nu$, then the span of the orbit of any function under C_φ has infinite codimension.*

Supercyclicity and hypercyclicity are extremely strong forms of cyclicity and, consequently, supercyclic or hypercyclic operators have many more properties than cyclic ones. For instance, a hypercyclic operator always has a residual set of hypercyclic vectors (see [Sh2 p. 109], for instance) and the same holds for supercyclic operators (see [Gr]). This is not always true for cyclic operators. An example of this behavior is the multiplication operator by z on $\mathcal{H}^2$. It is cyclic but it has no residual set of cyclic vectors. Therefore, it is interesting to know when an operator is supercyclic or it has a hypercyclic scalar multiple. The latter is the case of the first example of a hypercyclic operator in the Hilbert space setting that was due to Rolewicz [Ro]. He proved that if $B : \ell^2 \to \ell^2$ is the unweighted backward shift, then λB is hypercyclic for each complex number λ with $|\lambda| > 1$. Of course, B itself is not hypercyclic because it is a contraction. However, it is supercyclic. The concept of supercyclic operator was introduced by Hilden and Wallen [HW] in the mid seventies. In particular, they proved that unilateral weighted shifts are supercyclic operators on ℓ^2.

The following theorem shows that only a few linear fractional composition operators are cyclic on the Dirichlet space.

THEOREM 1.8. *Let φ be a non elliptic linear fractional map of $\mathbb{D}$. On the Dirichlet space $\mathcal{D}$ we have*
  (i) *The operator C_φ is cyclic if and only if φ is a hyperbolic or a parabolic non-automorphism or it has an exterior fixed point.*
  (ii) *No linear fractional composition operator is hypercyclic on $\mathcal{D}$.*
  (iii) *The operator λC_φ is hypercyclic for some complex number λ if and only if φ is a hyperbolic non-automorphism. In such a case, λC_φ is hypercyclic if and only if $|\lambda| > 1$.*
  (iv) *The operator C_φ is supercyclic if only if φ is a hyperbolic non automorphism.*

In this work we will also determine exactly the complex numbers λ for which λC_φ is hypercyclic. We will see that the solution to this problem is closely related to the spectrum of C_φ.

Table II below summarizes the hypercyclic behavior of λC_φ. In the cases not appearing in Table II, linear fractional maps induce composition operators that have no hypercyclic scalar multiple. There is hypercyclicity only when φ is a non-elliptic automorphism or a hyperbolic non automorphism. In any case, the attractive fixed point η is always on the boundary $\partial \mathbb{D}$. If φ is parabolic, then $\varphi'(\eta) = 1$ and if φ is hyperbolic, then $\varphi'(\eta) < 1$. The hypercyclicity of λC_φ depends on the value $\varphi'(\eta)$.

Type of φ	Hypercyclic	Example
Hyperbolic Automorphism	$\nu < 1/2$ and $\varphi'(\eta)^{(1-2\nu)/2} < \|\lambda\| < \varphi'(\eta)^{(2\nu-1)/2}$	$\dfrac{3z+1}{z+3}$
Parabolic Automorphism	$\nu < 1/2$ and $\|\lambda\| = 1$	$\dfrac{(1+i)z-1}{z+i-1}$
Hyperbolic Non-Automorphism	$\nu \leq 1/2$ and $\|\lambda\| > \varphi'(\eta)^{(1-2\nu)/2}$	$\dfrac{1+z}{2}$

TABLE II

If φ is a hyperbolic non automorphism, the spectrum of C_φ acting on $\mathcal{S}_\nu$ with $\nu \leq 1/2$ is the closed unit disk centered at the origin of radius $\varphi'(\eta)^{(2\nu-1)/2}$ (see [Hu, Thm. 8 and Cor. 12]). The point is that if $\nu > 1/2$ the spectrum of C_φ acting on $\mathcal{S}_\nu$ is (see [Hu, Cor. 12])

$$\{t : |t| \leq \varphi'(\eta)^{(2\nu-1)/2}\} \cup \{\varphi'(\eta)^j : j = 0, 1, \dots\}.$$

Thus from Table II we have the following Corollary.

COROLLARY 1.9. *Suppose that φ is a hyperbolic non-automorphism. Then*
(i) *If $\nu \leq 1/2$, the λC_φ is hypercyclic if and only if $1/\lambda$ is in the interior of the spectrum, that is, the unit circle is contained in the spectrum of λC_φ.*
(ii) *If $\nu > 1/2$, the λC_φ is never hypercyclic.*

Part (ii) follows immediately from a result of Kitai [Ki] that asserts that each component of the spectrum of a hypercyclic operator must meet the unit circle. Part (i) shows a limit case in which although the spectrum of λC_φ is the closed unit disk, that meets the unit circle, the operator is not hypercyclic.

Since the spectrum of C_φ, for φ a non elliptic automorphism, is unknown for $\nu > 0$, we could only state similar corollaries only when $\nu \leq 0$. For instance for the Bergman and the Hardy space. Thus, at the moment, a unified treatment from a spectral point of view is not possible. However, it is worth noting that the results on Table II suggest the picture that the spectrum should have.

We note that most of the results in the supercyclic column in Table I follow immediately from the fact that the supercyclicity is an intermediate property between cyclicity and hypercyclicity. In fact, if an operator is not cyclic, then it is not supercyclic either. On the other hand, as soon as an operator has a hypercyclic scalar multiple, then the operator is supercyclic. For instance, if φ is a parabolic non-automorphism, then C_φ is supercyclic on $\mathcal{D}$ because it has scalar multiples that are hypercyclic. At first glance, one might think that every supercyclic operator arises in this way. But this is not at all the case. In fact, as already mentioned unilateral weighted shifts are all supercyclic and some of them are also compact. Now, a result of Kitai [Ki] asserts that no compact operator can be hypercyclic. Therefore, no scalar multiple of a compact weighted shift can be hypercyclic. Moreover, Manuel González, Fernando León and the second named author have shown, that for $\alpha \neq 0$, the operator $S = \alpha I \oplus T : \mathbb{C} \oplus \mathcal{H} \to \mathbb{C} \oplus \mathcal{H}$ is supercyclic if and only if $(1/\alpha)T : \mathcal{H} \to \mathcal{H}$ is hypercyclic (see [GLM, Thm. 5.2]). On the other hand, since $\sigma_p(S^\star) = \{\alpha\} \neq \emptyset$, no scalar multiple of S can be hypercyclic [Ki, Cor. 2.4].

But the most interesting example is due to Salas [Sa]. This author has constructed an invertible supercyclic bilateral weighted shift such that any scalar multiple of it is not hypercyclic and has empty point spectrum. Supercyclic operators having empty point spectrum share many properties with hypercyclic operators (see [MS, Sections 1 and 2]).

We close the chapter with a more detailed outline of the following chapters.

Chapter 2 is devoted to studying the cyclic and hypercyclic behavior of linear fractional composition operators whose inducing symbols have an interior fixed point or an exterior fixed point. We will prove that if φ has an exterior fixed point, then C_φ is cyclic on $\mathcal{H}^2(\beta)$. The main point here is an argument of duality that allows us to use the reproducing kernels in the Hardy space (not in $\mathcal{H}^2(\beta)$). The cyclicity of C_φ will be shown for spaces with a slightly stronger hypothesis than boundedness of C_φ. We do not even require that $\mathcal{H}^2(\beta)$ be automorphism invariant. A weighted Hardy space is said to be automorphism invariant if the automorphisms of the unit disk induce bounded composition operators. Our results in this section extend and complement Zorboska's results [Zo3].

It will be also shown that these methods can be used to show that if φ is an elliptic automorphism that is conjugate to a rotation through an irrational multiple of π, then C_φ is cyclic on any $\mathcal{H}^2(\beta)$. The new thing here is that $\mathcal{H}^2(\beta)$ need not be automorphism invariant. The proof requires considerably more work than the corresponding proof for the Hardy space [BS2] and for the automorphism invariant case [Zo3].

It will be also proved that if φ has an interior and a boundary fixed point, then C_φ is strongly non cyclic on any of the $\mathcal{S}_\nu$ spaces. In this case, the method of proof is basically that of the corresponding result for the Hardy space that is due to Bourdon and Shapiro [BS2].

Finally, we end Chapter 2 with the study of the hypercyclicity of λC_φ on $\mathcal{S}_\nu$, where φ is a hyperbolic non-automorphism (the cyclicity is already dispensed). When $\mathcal{S}_\nu$ contains the Dirichlet space, the hypercyclicity of λC_φ depends on whether $1/\lambda$ belongs to the interior of the spectrum. If $\mathcal{S}_\nu$ is strictly smaller than the Dirichlet space, then λC_φ is not hypercyclic, but it is always hypercyclic on an invariant subspace of finite codimension. The cut-off will be determined by a Beurling type theorem about the density of some families of polynomials in the $\mathcal{S}_\nu$ spaces. In this section we also introduce a very simple and useful non hypercyclicity criterion that has interest in its own right. Indeed, it allows us to prove the non hypercyclicity of a given operator even when the spectrum meets the unit circle (a necessary condition for hypercyclicity that was discovered by Kitai [Ki]).

In Chapter 3 we study the case in which φ is a non elliptic automorphism of the unit disk. Again, the spectrum of C_φ will determine the hypercyclicity of λC_φ when the space $\mathcal{S}_\nu$ is strictly greater than the Dirichlet space. At least, this is true when the spectrum of C_φ is known, but that is not always the case. As for cyclicity, after discussing the best we can do using spectral methods, we will find, by a completely different approach that C_φ is strongly non cyclic on the Dirichlet space. For instance, for a hyperbolic automorphism the Haar measure with respect to the multiplicative group of positive real numbers is needed to prove that C_φ is not cyclic.

In Chapter 4 we study C_φ when the inducing symbol is a parabolic non-automorphism. The punch line of cyclicity will be the growth restrictions on the sequence of zeros of functions in $\mathcal{S}_\nu$ as well as the duality argument that we use in

Chapter 2. With respect to hypercyclicity we will prove that λC_φ is never hypercyclic on any of the $\mathcal{S}_\nu$ spaces. This is the corresponding result for the $\mathcal{S}_\nu$ spaces of an unpublished result of Shapiro for the Hardy space [Sh3]. This extension of Shapiro's result is not trivial. Indeed, the proof depends on a recursion argument that has its own interest. This latter argument, that involves functional Hilbert spaces and $L^2(\mathbb{R}^+, dt/t)$, will allow us to extend, via density, an orthogonal decomposition from one space to another.

In Chapter 5[1] we determine exactly which linear fractional composition operators are supercyclic. In this way, we extend a previous result of the authors for the Hardy space [GM1] to the $\mathcal{S}_\nu$ spaces (in particular to the Bergman space). The main problem here is to determine the non supercyclicity of C_φ where φ is a parabolic non automorphism. Although the fundamental ideas are those in [GM1], the techniques we use here are very different. The techniques in [GM1] are based on certain orthogonality properties of the eigenfunctions of C_φ that allowed us to study the behavior of C_φ acting on finite dimensional C_φ-invariant subspaces. It seems that these techniques do not generalize easily to the $\mathcal{S}_\nu$ spaces because the eigenfunctions do not have nice orthogonality properties as in $\mathcal{H}^2$. However, we can use the three term recurrence relation of the Laguerre polynomials and infinite matrices to estimate the behavior of the adjoint $C_\varphi^\star$ acting on certain C_φ-invariant subspaces.

Acknowledgment.

Before proceeding further we would like to thank Patrick R. Ahern, Daniel M. Burns, Carl C. Cowen, Peter L. Duren, Carmen Romero, Héctor N. Salas and Joel H. Shapiro for many helpful comments and suggestions.

---

[1]When this monograph was first submitted (February, 22, 2000) we only dealt with cyclic and hypercyclic scalar multiples of composition operators and not with supercyclic ones. The corresponding results about supercyclic operators were added in a later resubmission (February, 2001). In order to maintain the rest of the work unchanged we have included all the results about supercyclic linear fractional composition operators in Chapter 5. The main result about supercyclic operators that corresponds to Theorem 5.4 was obtained by the authors in September 2000 (an announcement without proofs of the results about supercyclicity can be found in [GM2]). Chapters 2 to 4 remain without any essential change from the original work.

CHAPTER 2

LINEAR FRACTIONAL TRANSFORMATIONS WITH AN INTERIOR OR AN EXTERIOR FIXED POINT

This chapter is devoted to studying the cyclicity of C_φ and the hypercyclicity of λC_φ when φ has an interior or an exterior fixed point. First we will dispense with the cyclicity. We will see that the cyclicity depends strongly on the location of the fixed points of φ. The hypercyclicity of λC_φ only will be possible when φ has no interior fixed point in $\mathbb{D}$.

First we consider the case in which φ has an exterior fixed point and is not an elliptic automorphism. We will establish the cyclicity of C_φ on the weighted Hardy spaces with a natural hypothesis related to the boundedness of C_φ. Then, we will determine the cyclicity of C_φ for φ an elliptic automorphism in weighted Hardy spaces requiring only the boundedness of C_φ. After that, it will be shown that if φ has an interior and a boundary fixed point, then it is not cyclic in a very strong sense. Finally, with respect to hypercyclicity, we will see that only composition operators induced by maps without interior fixed point may be hypercyclic, and the hypercyclicity of λC_φ with φ a hyperbolic non automorphism will be determined.

NON ELLIPTIC LINEAR FRACTIONAL MAPS WITH AN EXTERIOR FIXED POINT

By Proposition 1.4, the case in which φ is not elliptic and has an exterior fixed point comprises the case in which φ has an interior and an exterior fixed point and, the case in which φ has a boundary and exterior fixed point (the hyperbolic non-automorphism case). We will see that the cyclicity of C_φ depends essentially on the exterior fixed point.

Using Clancey-Rogers' Theorem as well as some special results on composition operators, Zorboska [Zo3, Prop. 4.1] has proved the cyclicity for bounded C_φ in the case that φ fixes the origin and the sequence of weights β is monotonically increasing. Our theorem lacks these restrictions. We will prove a general theorem about the cyclicity of C_φ in any $\mathcal{H}^2(\beta)$ with a very natural restriction that has to do with the boundedness of C_φ on $\mathcal{H}^2(\beta)$. Our proof is completely elementary and allows us to exhibit examples of cyclic vectors for C_φ. This provides a more penetrating proof which works equally well for the automorphism invariant case ([Zo3, Thm. 3.5]) as well as for the non-automorphism invariant case ([Zo3, Prop. 4.1]).

THEOREM 2.1. *Let $\mathcal{H}^2(\beta)$ be a weighted Hardy space. Assume that φ is a non-elliptic linear fractional self map of the unit disk with an exterior fixed point p and* induces a bounded composition operator on $\mathcal{H}^2(\beta)$. If $\limsup \beta_n^{1/n} < |p|$, then C_φ is cyclic on $\mathcal{H}^2(\beta)$. In particular, C_φ is cyclic on any of the $\mathcal{S}_\nu$ spaces.

REMARK 2.2. If $p \neq \infty$, the additional condition $\limsup \beta_n^{1/n} < |p|$ is a little bit more than a necessary condition for the operator C_φ to be bounded on $\mathcal{H}^2(\beta)$. In fact, in the course of the proof of Theorem 2.1, we will see that if C_φ is bounded on $\mathcal{H}^2(\beta)$, then $\limsup \beta_n^{1/n} \leq |p|$. We point out here that it is quite unusual to obtain a result on weighted Hardy spaces without imposing any hypothesis on the weights.

PROOF OF THEOREM 2.1. By Proposition 1.4 there must be another fixed point q in the closed unit disk, that is, with $|q| \leq 1$. In addition, q has to be the attractive fixed point of φ. Upon conjugating with

$$\sigma(z) = \frac{z-q}{z-p}$$

we have

$$w = \varphi(z) \equiv \frac{w-q}{w-p} = \mu \frac{z-q}{z-p} \qquad \text{with } 0 < |\mu| < 1.$$

Upon replacing μ by μ^n in the formula above we can easily get a formula for the iterates of φ

$$\varphi_n(z) = \frac{(q - \mu^n p)z + pq(\mu^n - 1)}{(1 - \mu^n)z + \mu^n q - p}. \tag{1}$$

Although, any cyclic property is invariant under similarity, we are not allowed to locate conveniently the fixed points because we are not assuming that $\mathcal{H}^2(\beta)$ is automorphism invariant.

We have to distinguish three cases. The common key point to these three cases that makes the proof work is that we can associate with any function $f(z) = \sum_{k=0}^{\infty} a_k z^k \in \mathcal{H}^2(\beta)$ the function

$$g(z) = \sum_{k=0}^{\infty} a_k \beta_k^2 z^k.$$

Now, we set $1/R = \limsup \beta_n^{1/n}$. Since $\sum_{n=0}^{\infty} |a_n|^2 \beta_n^2$ is a convergent series, we have $\limsup_{k \to \infty} (|a_n|\beta_n)^{1/n} \leq 1$. Therefore,

$$\limsup_{n \to \infty} (|a_n|\beta_n^2)^{1/n} \leq 1/R.$$

Thus all the functions g above are holomorphic, at least, on the disk $D(0, R)$. Also for $w \in D(0, R)$ the condition $1/R = \limsup \beta_n^{1/n}$ insures that the reproducing kernel (in the Hardy space) at the point w

$$K_w(z) = \frac{1}{1 - \bar{w}z} = \sum_{k=0}^{\infty} \bar{w}^k z^k$$

belongs to $\mathcal{H}^2(\beta)$. Thus we can compute

$$\langle f, K_w \rangle = \sum_{k=0}^{\infty} a_k \beta_k^2 w^k = g(w).$$

Case 1. $p \neq \infty$ and $q \neq 0$. We claim that the identity function $u(z) = z$, defined on $\mathbb{D}$, is a cyclic vector for C_φ. From (1), one easily obtains the decomposition

$$\varphi_n(z) = \bar{\gamma}_n + \bar{\alpha}_n K_{w_n}, \tag{2}$$

where

$$\bar{\gamma}_n = \frac{q - p\mu^n}{1 - \mu^n}, \quad \bar{\alpha}_n = \frac{(p-q)^2 \mu^n}{(1-\mu^n)(p - q\mu^n)} \quad \text{and} \quad \bar{w}_n = \frac{1 - \mu^n}{p - q\mu^n}. \tag{3}$$

The fact that $0 < |\mu| < 1$, $|p| > 1$ and $|q| \leq 1$ insures that the denominators above are always different from zero. Observe that since we are assuming that C_φ is bounded on $\mathcal{H}^2(\beta)$, then each φ_n must be in $\mathcal{H}^2(\beta)$. As a consequence, by (2), the reproducing kernel K_{w_n} must also be in $\mathcal{H}^2(\beta)$. Since the sequence $\{w_n\}$ tends to $1/\bar{p}$, we see that $\limsup \beta_n^{1/n} \leq |p|$. We point out here that the condition $\limsup \beta_n^{1/n} < |p|$ is to insure that $1/p \in D(0, R)$ and not to force $K_{w_n} \in \mathcal{H}^2(\beta)$.

Now suppose that $f \in \mathcal{H}^2(\beta)$ is orthogonal to the C_φ-orbit of u. To show the cyclicity of C_φ it is enough to prove that f is the zero function. First, we see that f is orthogonal to the constant functions. This is a consequence of the fact that $\{\varphi_n\}$ tends to $q \neq 0$ in the norm of $\mathcal{H}^2(\beta)$. In such a case we have

$$0 = \lim_{n \to \infty} \langle f, \varphi_n \rangle = \langle f, q \rangle = \bar{q} \beta_0^2 f(0).$$

Thus we have to prove that φ_n tends to q in the norm of $\mathcal{H}^2(\beta)$. Using the decomposition in (2) we have

$$\|\varphi_n - q\| \leq \|\bar{\gamma}_n - q\| + \|\bar{\alpha}_n K_{w_n}\| = \beta_0 |\bar{\gamma}_n - q| + |\alpha_n| \|K_{w_n}\|.$$

Since $\bar{\gamma}_n$ tends to q as n tends to ∞, it is enough to show $\|\bar{\alpha}_n K_{w_n}\| \to 0$. Now, $|w_n|$ tends to $1/|p| < R$. Thus for n large enough $|w_n| \leq r$ for some $r < R$. Hence, we can estimate

$$|\alpha_n|^2 \|K_{w_n}\|^2 = |\alpha_n|^2 \sum_{k=0}^{\infty} |w_n|^{2k} \beta_k^2 < |\alpha_n|^2 \sum_{k=0}^{\infty} r^{2k} \beta_k^2 \leq |\alpha_n|^2 \|K_r\|^2$$

that goes to zero because $\{\alpha_n\}$ tends to zero as n tends to ∞.

The fact that $f(0) = 0$ along with the orthogonality of f and φ_n implies

$$0 = \langle f, \varphi_n \rangle = \gamma_n \langle f, 1 \rangle + \alpha_n \langle f, K_{w_n} \rangle = \alpha_n \sum_{k=0}^{\infty} a_k w_n^k \beta_k^2 = \alpha_n g(w_n).$$

Observe that for n large enough $w_n \in D(0, R)$. Since $\alpha_n \neq 0$ for every n, we find that the function $g(w_n) = 0$ whenever n is large enough. Thus g vanishes on a sequence of points with a limit point in $D(0, R)$. This implies that g vanishes identically on $D(0, R)$. Hence, all the coefficients in the Taylor development of g are zero, and, consequently, f is the zero function. Therefore, it follows that u is a cyclic vector for C_φ.

Case 2. $p = \infty$. The fact that $1/R = \limsup \beta_n^{1/n} < \infty$ insures that $R > 0$. This time $u(z) = z$ is not a cyclic vector. However, we will show that for any $\alpha \in D(0, R) \setminus \{0\}$ the reproducing kernel (in the Hardy space) K_α is a cyclic vector for C_φ. In this case the iterates are

$$\varphi_n(z) = \mu^n z + q(1 - \mu^n) \quad \text{with } 0 < |\mu| < 1.$$

An easy computation yields
$$K_\alpha \circ \varphi_n(z) = \bar{a}_n K_{w_n},$$
where
$$\bar{a}_n = \frac{1}{1 - \bar{\alpha}q(1-\mu^n)} \quad \text{and} \quad \bar{w}_n = \frac{\bar{\alpha}\mu^n}{1 - \bar{\alpha}q(1-\mu^n)}.$$
This time the conditions on α, q and μ only imply that a_n and w_n are well defined except possibly for one fixed positive integer n_0. But this is enough to make the proof work. Let $f \in \mathcal{H}^2(\beta)$ be orthogonal to $\{K_\alpha \circ \varphi_n\}$. Again we will prove that f is the zero function. This time we do not need to prove that $f(0) = 0$. The orthogonality hypothesis on f implies, whenever $n \neq n_0$, that
$$\langle f, K_\alpha \circ \varphi_n \rangle = a_n \langle f, K_{w_n} \rangle = a_n g(w_n).$$
Thus the function g vanishes on the sequence $\{w_n\}_{n \neq n_0}$. Since $w_n \to 0$, the function g vanishes on a sequence with a limit point in $D(0, R)$. Thus we can conclude as in Case 1 that f is the zero function. This proves that K_α is a cyclic vector for C_φ.

Case 3. $q = 0$ and $p \neq \infty$. This case exhausts all the possibilities. Again, the function $u(z) = z$ is not cyclic because any vector in $\overline{\text{span}}\{\varphi_n : n \geq 0\}$ vanishes at 0. As in Case 2 we will prove that whenever $\alpha \in D(0, R) \setminus \{0, 1/\bar{p}\}$ the reproducing kernel K_α is a cyclic vector for C_φ. This time the formula for the iterates is given by
$$\varphi_n(z) = \frac{\mu^n p z}{p - (1-\mu^n)z} \quad \text{with } 0 < |\mu| < 1.$$
Now we have the decomposition
$$K_\alpha \circ \varphi_n = \bar{\gamma}_n + \bar{a}_n K_{w_n}, \qquad (4)$$
where
$$\bar{\gamma}_n = \frac{1 - \mu^n}{1 - \mu^n + \bar{\alpha}p\mu^n}, \quad \bar{a}_n = \frac{\bar{\alpha}\mu^n p}{1 - \mu^n + \bar{\alpha}p\mu^n} \quad \text{and} \quad \bar{w}_n = \frac{1 - \mu^n + \bar{\alpha}p\mu^n}{p}.$$
Again all the denominators above are different from zero except possibly for a positive integer n_0.

Suppose that f is orthogonal to the orbit $\{K_\alpha \circ \varphi_n\}$. We need to show that f is the zero function. First we see that f is orthogonal to the constant functions. This is a consequence of the fact that $K_\alpha \circ \varphi_n$ converges to 1 in the norm of $\mathcal{H}^2(\beta)$. Indeed, using the decomposition in (4) for $n \neq n_0$ we have
$$\|\bar{\gamma}_n + a_n K_{w_n} - 1\| \leq |\bar{\gamma}_n - 1|\beta_0 + |\bar{a}_n|\|K_{w_n}\|.$$
Since γ_n tends to 1, it is enough to show that the second term in the second member above tends to zero. But this follows as in Case 1 because $\{a_n\}$ tends to zero and $\{\bar{w}_n\}$ tends to $1/p \in D(0, R)$. Thus $f(0) = 0$.

The orthogonality condition of f and $K_\alpha \circ \varphi_n$ implies for $n \neq n_0$
$$0 = \langle f, K_\alpha \circ \varphi_n \rangle = \gamma_n \langle f, 1 \rangle + a_n \langle f, K_{w_n} \rangle = a_n g(w_n).$$
Since a_n is different from zero for every n, we may conclude that g vanishes on each element of the sequence $\{w_n\}_{n \neq n_0}$. As $\alpha \neq 1/\bar{p}$, the sequence $\{w_n\}$ is not eventually constant and has $1/\bar{p}$ as a limit point in $D(0, R)$. Therefore, as in the previous two cases, we can conclude that f is the zero function and that K_α is a cyclic vector for C_φ. The proof of the theorem is finished. □

ELLIPTIC AUTOMORPHISM

Now, we will study the cyclicity of C_φ on $\mathcal{H}^2(\beta)$ when φ is an elliptic automorphism of the unit disk. We will not assume that $\mathcal{H}^2(\beta)$ is automorphism invariant nor any additional hypothesis on the sequence of weights. By Proposition 1.4, an elliptic automorphism has an interior and an exterior fixed point. Again, the cyclicity of C_φ (when cyclic) depends on the exterior fixed point. The behavior of C_φ is like the limit case of the interior and exterior fixed point. As a consequence of this fact, a more complicated proof will be required (at least, when the exterior fixed point is not ∞). Indeed, the method of the preceding section must be refined. We have the following

THEOREM 2.3. *Let $\mathcal{H}^2(\beta)$ be a weighted Hardy space. Assume that φ is an elliptic automorphism of the unit disk and induces a bounded composition operator on $\mathcal{H}^2(\beta)$. Then C_φ is cyclic if and only if φ is conjugate to a rotation through an irrational multiple of π. In particular, the same is true for any of the $\mathcal{S}_\nu$ spaces.*

REMARK 2.4. For the Hardy space $\mathcal{H}^2$ the assertion of the above theorem is due to Bourdon and Shapiro (see [BS2, Prop. 2.1]) and Zorboska [Zo3] observed that the same proof works for automorphism invariant weighted Hardy spaces. Thus our contribution is to remove the automorphism invariant hypothesis and to exhibit explicitly cyclic vectors for C_φ.

PROOF OF THEOREM 2.3. Let p denote the exterior fixed point and let q denote the interior fixed point. As in the exterior fixed point case, by conjugating with

$$\sigma(z) = \frac{z-q}{z-p}$$

we get the following formula for the iterates of φ

$$\varphi_n(z) = \frac{(q - \mu^n p)z + pq(\mu^n - 1)}{(1 - \mu^n)z + \mu^n q - p} \qquad \text{with } |\mu| = 1. \tag{5}$$

The only difference from formula (1) is that, in the present situation, μ must be of modulus one. Now, it is clear from the expression of the iterates that if μ is a rational multiple of π, then C_φ fails to be cyclic because the orbit of any function under C_φ is a finite set.

Now, suppose that μ is an irrational multiple of π. This time we only need to distinguish two cases.

Case 1. $p = \infty$. Formula (5) becomes $\varphi(z) = \mu z + q - \mu q$ with $|q| < 1$. In this case q must be 0. If not, we may assume that q is real and since $|\varphi(1)|^2 = |\mu(1-q)+q|^2 = 1$, we have

$$2q(1 - \Re\mu)(q - 1) = 0.$$

It follows that $\Re\mu = 1$. Thus $\mu = 1$ and φ is the identity, a contradiction. Thus $q = 0$ and $\varphi(z) = \mu z$. We can proceed as in [BS2, Prop. 2.1]. In the present case, the identity function $u(z) = z$ is not a cyclic vector for C_φ. However, the reproducing kernel (this time in $\mathcal{H}^2(\beta)$)

$$K_\alpha^\beta(z) = \sum_{k=0}^\infty \frac{\bar{\alpha}^k}{\beta_k^2} z^k$$

at any point $\alpha \neq 0$ in the unit disk is a cyclic vector for C_φ. To show this, we suppose that f is orthogonal to the orbit $\{C_{\mu z}^n K_\alpha^\beta\} = \{K_{\bar\mu^n \alpha}^\beta\}$. Then for any non negative integer we have

$$0 = \langle f, K_{\bar\mu^n \alpha}\rangle = f(\bar\mu^n \alpha).$$

Hence, f must vanish at infinitely many points on the circle $|z| = \alpha$. Therefore, f must vanish identically on $\mathbb{D}$ and it follows that K_α^β is cyclic for $C_{\mu z}$.

Case 2. $p \neq \infty$. We claim that the identity function $u(z) = z$ is a cyclic vector for C_φ. From (5) we have

$$\varphi_n = \bar\gamma_n + \bar\alpha_n K_{w_n},$$

where $\bar\gamma_n$, $\bar\alpha_n$ and $\bar w_n$ satisfy formulae (3) with $|\mu| = 1$. Observe that these quantities are always well defined for any positive integer. Again, let K_w be the reproducing kernel at w in the Hardy space.

Let $f \in \mathcal{H}^2(\beta)$ be orthogonal to $\{\varphi_n\}$. We will show that f is the zero function. First we will show that f is orthogonal to the constant functions. Due to the fact that $|\mu| = 1$, this takes more work than that of the proof of Theorem 2.1.

To begin with we consider the function

$$\bar w(t) = \frac{1-t}{p-qt} \qquad t \in \partial\mathbb{D}.$$

Observe that $\bar w(\mu^n) = \bar w_n$ for every n. An exercise, involving the fact that μ is an irrational multiple of π, shows that $\{\mu^n\}$ is dense in the unit circle. Therefore, for any $t \in \partial\mathbb{D}$, we can extract a subsequence $\{n_j\}$ of positive integers such that $\mu^{n_j} \to t$ as $j \to \infty$. On the other hand, for $t = -1$, we have

$$|\bar w(-1)| = \frac{2}{|p+q|} > \frac{1}{|p|}.$$

Hence, by just taking a subsequence $\{n_j\}$ such that $\mu^{n_j} \to -1$, we see that there is n_0 such that $|\bar w_{n_0}| = |\bar w(\mu^{n_0})| > 1/|p|$. Now, since C_φ is bounded, so is C_{φ_n} for every n, and the decomposition in (1) shows that K_{w_n} belongs to $\mathcal{H}^2(\beta)$ for every n. In particular, $K_{w_{n_0}}$ is in $\mathcal{H}^2(\beta)$. This forces to have $1/R = \limsup \beta_n^{1/n} < |p|$. Thus in the same way as in the proof of Theorem 2.1 we may associate to each function $h(z) = \sum_{k=0}^\infty a_k z^k \in \mathcal{H}^2(\beta)$ the function

$$g(z) = \sum_{k=0}^\infty a_k \beta_k^2 z^k$$

which is holomorphic on the disk $D(0, R)$. We observe that K_w belongs to $\mathcal{H}^2(\beta)$ whenever $|w| < R$.

Now, fix $0 < r < R$, and observe that $\bar w(1) = 0$. Therefore, by continuity, there is a non degenerate interval L on the unit circle such that

$$|\bar w(t)| < r \quad \text{for } t \in L.$$

Obviously, we may suppose that 1 is not in L.

For the moment fix $t \in L$ and choose a sequence $\{n_j\}$ of positive integers such that $\{\mu^{n_j}\}$ tends to t as $j \to \infty$. Consequently, $\bar\gamma_{n_j} \to \bar\gamma$, $\bar\alpha_{n_j} \to \bar\alpha$ and $\bar w_{n_j} \to \bar w$ where

$$\bar\gamma = \frac{q-pt}{1-t}, \quad \bar\alpha = \frac{(p-q)^2 t}{(1-t)(p-qt)} \quad \text{and} \quad \bar w = \frac{1-t}{p-qt}.$$

Also, it is clear that $\{\varphi_{n_j}\}$ tends to $\bar{\gamma} + \bar{\alpha} K_w$ uniformly on compact subsets of $\mathbb{D}$. In addition, we can show that $\{\varphi_{n_j}\}$ remains bounded in the norm of $\mathcal{H}^2(\beta)$. To see this, we note that $\{\bar{w}_{n_j}\}$ converges to $\bar{w}$ and $|\bar{w}| < r$. Thus for j large enough, we have $|w_{n_j}| \leq r$ and hence we may estimate

$$\|\varphi_{n_j}\| = \|\bar{\gamma}_{n_j} + \bar{\alpha}_{n_j} K_{w_{n_j}}\| \leq (|\gamma|+1)\beta_0 + (|\alpha|+1)\|K_r\|.$$

Thus we may conclude that $\{\varphi_{n_j}\}$ tends weakly to $\bar{\gamma} + \bar{\alpha} K_w$. This along with the hypothesis of orthogonality of f and φ_n yields

$$0 = \lim_{j\to\infty} \langle f, \varphi_{n_j}\rangle = \gamma\langle f, 1\rangle + \alpha\langle f, K_w\rangle = \beta_0^2 \gamma f(0) + \alpha g(w).$$

Since $\alpha \neq 0$, we have $g(w) = -(\gamma/\alpha)\beta_0^2 f(0)$. On the other hand, since $w = (1-\bar{t})/(\bar{p}-\bar{q}\bar{t})$ and $|t|=1$, we have

$$t = \frac{\bar{q}w - 1}{\bar{p}w - 1}.$$

Upon substituting the value of t in the expressions of γ and α we obtain

$$g(w) = -\frac{\gamma}{\alpha}\beta_0^2 f(0) = \frac{1-(\bar{p}+\bar{q})w}{(w\bar{p}-1)(w\bar{q}-1)}\beta_0^2 f(0). \tag{6}$$

Letting t vary along L we see that the expression above must be true for the set

$$\left\{w(t): \bar{w}(t) = \frac{1-t}{p-qt} \text{ and } t \in L\right\}$$

which is contained in $D(0,R)$. The analytic identity principle implies that equality (6) must be true for the whole disk $D(0,R)$. But, for $f(0) \neq 0$ this is impossible because the right hand side in (6) has a pole at $w = 1/\bar{p} \in D(0,R)$ and g is analytic on this disk. Thus we may conclude that f is orthogonal to the constant functions.

To finish the proof, fix $t \in L$ again. Then, there is a subsequence $\{n_j\}$ such that μ^{n_j} tends to t and $\bar{w}_{n_j}$ tends to $\bar{w}(t)$. In particular, $|w_{n_j}| < R$ for j large enough. Thus g is well defined on w_{n_j} for j large enough. We have

$$0 = \langle f, \varphi_{n_j}\rangle = \langle f, \bar{\gamma}_{n_j}\rangle + \alpha_{n_j}\langle f, K_{w_{n_j}}\rangle = \alpha_{n_j} g(w_{n_j}).$$

Hence, g vanishes on a sequence with a limit point in $D(0,R)$. Consequently, g is the null function and, therefore, so is f. It follows that u is a cyclic vector for C_φ and the proof of the theorem is finished. $\square$

REMARK 2.5. We have seen that when C_φ fails to be cyclic the span of the orbit of any function is finite dimensional and, consequently, C_φ is strongly non cyclic.

INTERIOR AND BOUNDARY FIXED POINT

In this section we deal with the case in which φ has an interior and a boundary fixed point. We will see that in this case C_φ is strongly non cyclic on any of the $\mathcal{S}_\nu$ spaces.

THEOREM 2.6. *Let φ be a linear fractional self map of the unit disk with an interior fixed point and a boundary fixed point. Then C_φ is not cyclic on any of the spaces $\mathcal{S}_\nu$; in fact, the closed linear span of the orbit of any vector under C_φ has infinite codimension in $\mathcal{S}_\nu$.*

To prove Theorem 2.6 we need the equivalent norm in $\mathcal{S}_\nu$ furnished by Lemma 1.2. Recall that $dA(z) = (1/\pi)dxdy = (1/\pi)rdrd\theta$, $(z = x+iy = re^{i\theta})$ denotes the normalized Lebesgue area measure on $\mathbb{D}$. For $\nu < 0$ we take $l = -1$ in Lemma 1.2 and set $c = -1 - 2\nu$. Thus we have the following equivalent norm in $\mathcal{S}_\nu$

$$\|f\|^2 = \int_{\mathbb{D}} |f(z)|^2 (1-|z|^2)^c \, dA(z) \tag{7}$$

whenever $\nu < 0$.

PROOF OF THEOREM 2.6. If φ is the identity map, then the result is trivial because C_φ is the identity operator. Suppose that φ is not the identity map. Without loss of generality we may suppose that the fixed points of φ are 0 and 1. Upon conjugating with

$$\sigma(z) = \frac{1+z}{1-z}$$

that takes the unit disk onto the right half plane we can see that φ satisfies the formula

$$\varphi(z) = \frac{\mu z}{1 - (1-\mu)z} \quad \text{with} \quad 0 < \mu < 1.$$

By the Comparison Principle it is enough to prove the result for $\nu < 0$. Let $\mathcal{S}_\nu^0$ denote the closed subspace consisting of $\mathcal{S}_\nu$ functions that vanish at the origin. Now, since φ fixes the origin, the space $\mathcal{S}_\nu^0$ is invariant under C_φ. In addition, the space of constant functions is invariant under any composition operator. Since $\mathcal{S}_\nu$ is the orthogonal sum of both subspaces, they are also invariant for the adjoint C_φ^*.

The method of the proof will consist of computing the adjoint C_φ^* of C_φ acting on $\mathcal{S}_\nu^0$. We will find that this adjoint is similar to a composition operator acting on $\mathcal{S}_{-\nu}^0$. Then, since $\mathcal{S}_{-\nu}$ is contained in the Hardy space, we can proceed in a similar way as in [BS2, Thm. 2.8].

Let $f(z) = \sum_{k=1}^\infty a_k z^k$ be any function in $\mathcal{S}_\nu^0$. Using the norm in (7) we compute the norm of the monomials is

$$\begin{aligned} w_k &= \|z^k\| \\ &= \left(\int_{\mathbb{D}} |z|^{2k} (1-|z|^2)^c \, dA(z) \right)^{1/2} \\ &= \left(2 \int_0^1 r^{2k+1} (1-r^2)^c \, dr \right)^{1/2}. \end{aligned}$$

Set $u_n(z) = z^n/w_n$. Now if n is any positive integer we have

$$\begin{aligned} \langle C_\varphi^* f, u_n \rangle &= \langle f, C_\varphi u_n \rangle \\ &= \langle f, u_n \circ \varphi \rangle \\ &= \frac{1}{w_n} \langle f, \varphi^n \rangle \\ &= \frac{1}{w_n} \int_{\mathbb{D}} f(z) \overline{\varphi(z)}^n (1-|z|^2)^c \, dA(z). \end{aligned} \tag{8}$$

The orthogonality of the monomials with respect to $(1-|z|^2)dA(z)$ along with the formula
$$\frac{1}{(1-t)^n} = \sum_{k=n}^{\infty} \binom{k-1}{n-1} t^{k-n}$$
allows us to compute
$$\int_{\mathbb{D}} f(z) \left[\frac{\mu \bar{z}}{1-(1-\mu)\bar{z}}\right]^n (1-|z|^2)^c dA(z)$$
$$= 2\int_0^1 \sum_{k=n}^{\infty} \binom{k-1}{n-1} a_k \mu^n (1-\mu)^{k-n} r^{2k+1} (1-r^2)^c dr$$
$$= \mu^n \sum_{k=n}^{\infty} \binom{k-1}{n-1} a_k w_k^2 (1-\mu)^{k-n}. \tag{9}$$

We consider the diagonal operator on $\mathcal{S}_\nu^0$ defined by
$$Wf(z) = \sum_{k=1}^{\infty} a_k w_k^2 z^k \qquad (Wu_k = w_k^2 u_k).$$

Since $\nu < 0$, the sequence $\{w_k^2\}$ is bounded and, therefore, so is W. Indeed, it is obvious that W defines a unitary isometry between $\mathcal{S}_\nu^0$ and $\mathcal{S}_{-\nu}^0$ (to be precise here, $\mathcal{S}_{-\nu}$ must be considered with an equivalent norm). Therefore, the range of W is contained in $\mathcal{H}_0^2$, the subspace of Hardy functions that vanish at the origin. To save notation we write $g(z) = Wf(z) \in \mathcal{H}_0^2$. We also set $h(z) = g(z)/z$ and $\psi(z) = \mu z + 1 - \mu$. Since g is in the Hardy space, we can recover the integral from (9) to get
$$\frac{\mu^n}{2\pi} \int_0^{2\pi} g(z) \left[\frac{\bar{z}}{1-(1-\mu)\bar{z}}\right]^n d\theta = \frac{\mu^n}{2\pi i} \int_{|z|=1} \frac{h(z)}{[z-(1-\mu)]^n} dz \tag{10}$$

Now, using the Cauchy integral formula for the derivative in the second equality below we see that (10) is equal to
$$\frac{\mu^n}{(n-1)!} h^{n-1)}(1-\mu) = \frac{\mu}{(n-1)!} (h \circ \psi)^{n-1)}(0).$$

Let M_z and $M_{1/z}$ denote respectively the operator of multiplication by z on $\mathcal{S}_\nu$, and the multiplication by $1/z$ on $\mathcal{S}_{-\nu}^0$. Observe that the last quantity is the $(n-1)$-st Taylor coefficient in the development of $\mu h \circ \psi$ around the origin. Since C_ψ is bounded on $\mathcal{S}_{-\nu}$, it takes $\mathcal{S}_{-\nu}$ into itself. Thus, upon substituting in (8) we have
$$\langle C_\varphi^\star f, u_n \rangle = \frac{\mu}{w_{n-1} w_n} \langle h \circ \psi, u_{n-1} \rangle$$
$$= \frac{\mu}{w_{n-1} w_n} \langle C_\psi M_{1/z} Wf, u_{n-1} \rangle$$
$$= \frac{\mu}{w_n^2} \langle M_z C_\psi M_{1/z} Wf, u_n \rangle$$
$$= \mu \langle W^{-1} M_z C_\psi M_{1/z} Wf, u_n \rangle,$$
where the inner product was always that of $\mathcal{S}_\nu$. Thus we have shown
$$C_\varphi^\star|_{\mathcal{S}_\nu^0} = \mu W^{-1} M_z C_\psi M_{1/z} W.$$

It is easy to see that each function $(1-z)^\lambda$ for $\Re\lambda > -1/2$ is in $\mathcal{H}^2$ and, in addition, each of these functions is an eigenvector for C_ψ in $\mathcal{H}^2$ corresponding to the eigenvalue μ^λ. On the other hand, if $\Re\lambda > -\nu - 1/2$, then

$$\sum_{n=0}^{\infty} \left| \binom{\lambda}{n} \right|^2 (n+1)^{-2\nu} < \infty$$

and, therefore, $(1-z)^\lambda \in \mathcal{S}_{-\nu}$. We conclude that the function

$$f_\lambda(z) = W^{-1}(z(1-z)^\lambda)$$

is in $\mathcal{S}_\nu$. Thus by the formula for the adjoint of C_φ

$$C_\varphi^\star f_\lambda = \mu^{\lambda+1} f_\lambda.$$

Now, we fix λ, with $\Re\lambda > -\nu - 1/2$. Then, one checks easily that the set of Hardy functions

$$\{(1-z)^{\lambda(k)} : \lambda(k) = \lambda + 2\pi i k / \log \mu : k \in \mathbb{Z}\}$$

is linearly independent and, therefore, $\{f_{\lambda(k)} : k \in \mathbb{Z}\}$ is also linearly independent. Finally, we observe that

$$C_\varphi^\star f_{\lambda(k)} = \mu^{\lambda(k)+1} f_{\lambda(k)} = \mu^{\lambda+1} f_{\lambda(k)}.$$

Therefore, $\mu^{\lambda+1}$ is an eigenvalue for $C_\varphi^\star$ with infinite multiplicity. Now, the statement of the theorem follows from a result of Bourdon and Shapiro [BS2, Prop. 2.7] that asserts that if the adjoint of an operator has an eigenvalue of infinite multiplicity, then the span of the orbit of any vector has infinite codimension. □

REMARK 2.7. Hurst [Hu, Thm. 5] has computed the spectrum of C_φ in the $\mathcal{S}_\nu$ spaces when φ is a non-parabolic linear fractional map that has exactly one boundary fixed point and we could have used his result. But, besides the sake of completeness, the way we have obtained the expression for the adjoint in the particular case of interior and boundary fixed point is different from that of Hurst and similar to that of Bourdon and Shapiro.

REMARK 2.8. Theorem 2.6 along with the Comparison Principle allows us to determine that C_φ is not cyclic in many of the spaces $\mathcal{H}^2(\beta)$ in which C_φ is bounded, even if these latter spaces are not automorphism invariant. In fact, in some sense Theorem 2.6 is the best possible. For instance, suppose that the sequence of weights satisfies $\lim_{n\to\infty} n^\nu \beta_n = 0$ for every ν. In this case, the space $\mathcal{H}^2(\beta)$ contains any of the spaces $\mathcal{S}_\nu$. But in this situation, Theorem 5.2 in [CM] asserts that C_φ is not bounded whenever there is a point $\xi \in \partial \mathbb{D}$ such that $|\varphi'(\xi)| < 1$. This latter fact is readily seen as soon as one takes into account that if φ is as in Theorem 2.6 then it applies $\overline{\mathbb{D}}$ onto a smaller disk.

On the other hand, if we consider the whole space of analytic functions $\mathcal{H}(\mathbb{D})$ endowed with the topology of the uniform convergence on compacta the operator C_φ becomes cyclic. Indeed, we have the following

THEOREM 2.9. *Let φ be a linear fractional self map of the unit disk with an interior fixed point and a boundary fixed point. Then C_φ is cyclic on $\mathcal{H}(\mathbb{D})$.*

PROOF. As in the proof of Theorem 2.1 one can easily get the following formula for the iterates of φ

$$\varphi_n(z) = \frac{\mu^n p z}{p - (1-\mu^n)z} \quad \text{with } 0 < |\mu| < 1,$$

where we have supposed that $q = 0$ is the interior fixed point and p is the boundary fixed point. Let K_α be the reproducing kernel at the Hardy space at the point $\alpha \in \mathbb{D} \setminus \{0\}$. We will show that K_α is a cyclic vector for C_φ. As in case 3 in Theorem 2.1, we have

$$K_\alpha \circ \varphi_n = \bar{\gamma}_n + \bar{\alpha}_n K_{w_n},$$

where

$$\bar{\gamma}_n = \frac{1-\mu^n}{1-\mu^n + \bar{\alpha}p\mu^n}, \quad \bar{\alpha}_n = \frac{\bar{\alpha}\mu^n p}{1-\mu^n + \bar{\alpha}p\mu^n} \quad \text{and} \quad \bar{w}_n = \frac{1-\mu^n + \bar{\alpha}p\mu^n}{p}.$$

The quantities above are all well defined except possibly for a positive integer n_0.

Let f be in the dual space of $\mathcal{H}(\mathbb{D})$. Then f must be analytic on the exterior of a closed disk $D(0,r)$ with $r < 1$ and $f(\infty) = 0$ (see [Co, Prop. 4.2]). Thus $f(z) = \sum_{n=1}^\infty b_n (1/z)^n$ with $R = \limsup |b_n|^{1/n} < 1$. For $h(z) = \sum_{n=0}^\infty a_n z^n \in \mathcal{H}(\mathbb{D})$, the duality is realized under the pairing

$$\langle f, h \rangle = \sum_{n=0}^\infty a_n b_{n+1}.$$

Now, suppose that f as a linear functional on $\mathcal{H}(\mathbb{D})$ vanishes on the orbit $\{K_\alpha \circ \varphi_n\}$. To prove that K_α is a C_φ-cyclic vector, it is enough to prove that f is the zero function. Since $K_\alpha \circ \varphi_n$ tends to 1 uniformly on compact subsets of $\mathbb{D}$, it follows that $\langle f, 1 \rangle = 0$. Therefore, for $n \neq n_0$

$$0 = \langle f, K_\alpha \circ \varphi_n \rangle = \gamma_n \langle f, 1 \rangle + \alpha_n \langle f, K_{w_n} \rangle = \alpha_n g(\bar{w}_n),$$

where $g(z) = \sum_{n=0}^\infty b_{n+1} z^n$ is analytic on the disk $D(0, 1/R)$. Since $\alpha_n \neq 0$ for $n = 1, 2, \ldots$, we find $g(\bar{w}_n) = 0$ for $n = 1, 2, \ldots$ and $n \neq n_0$. Furthermore, since $|\alpha| < 1$, the sequence $\{w_n\}$ is not eventually constant and has $1/\bar{p}$, which has modulus 1, as a limit point in $D(0, 1/R)$. Finally, the identity principle shows that g is the zero function and therefore, so is f. The proof is finished. □

HYPERCYCLICITY

In this section, we first show that the hypercyclicity of λC_φ is impossible if φ has an interior fixed point. Then we will turn our attention to the hyperbolic non automorphism case.

PROPOSITION 2.10. *Let φ be a holomorphic self map of $\mathbb{D}$ with an interior fixed point on $\mathbb{D}$. Suppose that C_φ is bounded on $\mathcal{H}^2(\beta)$. Then for each $\lambda \in \mathbb{C}$, the operator λC_φ is not hypercyclic on $\mathcal{H}^2(\beta)$.*

PROOF. Let $p \in \mathbb{D}$ be the fixed point of φ. Suppose that f is hypercyclic for λC_φ and suppose that for $g \in \mathcal{H}^2(\beta)$ there is a sequence $\{n_k\}$ such that $\lambda^{n_k} C_\varphi^{n_k} f$ tends to g in $\mathcal{H}^2(\beta)$. Since norm convergence in $\mathcal{H}^2(\beta)$ implies pointwise convergence, it follows that

$$g(p) = \lim_k \lambda^{n_k} (C_{\varphi_{n_k}} f)(p) = \lim_k \lambda^{n_k} f(\varphi_{n_k}(p)).$$

Now, observe that $f(\varphi_{n_k}(p)) \to f(p)$ as $k \to \infty$. Therefore, if $|\lambda| < 1$, then $g(p) = 0$ that is not the case for every function $g \in \mathcal{H}^2(\beta)$. If $|\lambda| = 1$, then $|g(p)| = |f(p)|$ that is neither the case for every $g \in \mathcal{H}^2(\beta)$. Finally, if $|\lambda| > 1$, then $g(p)$ is not even defined, unless (and not always) $f(p) = 0$. But $f(p)$ cannot be zero for every hypercyclic vector, because the set of hypercyclic vectors is a residual dense subset. Thus in any case the orbit of f under λC_φ cannot be dense in $\mathcal{H}^2(\beta)$. □

Indeed, under the hypotheses of Proposition 2.10, C_φ is not even supercyclic. This will be seen in Chapter 5, Theorem 5.2.

It remains to analyze the hypercyclicity of λC_φ in the case that φ has a boundary fixed point. By Proposition 1.4, φ must be a hyperbolic non automorphism. In this case, the hypercyclicity of λC_φ is basically determined by a Beurling type Theorem about density of certain subsets of $\mathcal{S}_\nu$.

Hyperbolic non automorphism.

If φ is a hyperbolic non automorphism, we know that φ must have a boundary fixed point and an exterior fixed point. Therefore, we already know, by Theorem 2.1, that the corresponding composition operator C_φ acting on any of the $\mathcal{S}_\nu$ spaces is always cyclic. Thus it remains to study the hypercyclicity of λC_φ. We will see that the cut-off of hypercyclicity is quite mild. In fact, we can always find a complex number λ such that λC_φ is hypercyclic on an invariant subspace of finite codimension.

It is easy to see that if η is the boundary fixed point, then $0 < \varphi'(\eta) < 1$. With this observation in mind we can state our main result in this section.

THEOREM 2.11. *Let φ be a hyperbolic non-automorphism and η its boundary fixed point. Then λC_φ is hypercyclic on $\mathcal{S}_\nu$ if and only if $|\lambda| > \varphi'(\eta)^{(1-2\nu)/2}$ and $\nu \leq 1/2$. In particular, we have*
  (i) *For the Dirichlet space ($\nu = 1/2$) the operator λC_φ is hypercyclic if and only if $|\lambda| > 1$.*
  (ii) *The operator C_φ is hypercyclic on $\mathcal{S}_\nu$ if and only if $\nu < 1/2$.*

REMARK 2.12. Parts (i) and (ii) follow immediately. We have stated them to highlight the results. The hypercyclicity of C_φ was also stated in [Zo3, Prop. 3.4] but there is an oversight because it is asserted that C_φ is hypercyclic for $\nu = 1/2$, the Dirichlet space. It is worth mentioning that the hypercyclic behavior of C_φ acting on the Dirichlet space is exactly the same as the hypercyclic behavior of the unweighted backward shift acting on the sequence space ℓ^2. In fact, the spectrum of both operators is the closed unit disk.

The Hypercyclicity Criterion.

To prove Theorem 2.11 we will use the Hypercyclicity Criterion that has become well known in the literature in the field (see [Sh2, Chap. 7], for instance).

THEOREM 2.12 (Hypercyclicity Criterion). *Let T be a bounded linear operator on a separable Hilbert space $\mathcal{H}$. Suppose that there are*
  (a) *A dense subset $X \subset \mathcal{H}$ such that $\|T^n x\| \to 0$ for every $x \in X$.*
  (b) *A dense subset $Y \subset \mathcal{H}$ and a mapping $S : Y \to Y$ such that $TS =$ identity on Y, and $\|S^n y\| \to 0$ for every $y \in Y$.*
Then T is hypercyclic.

Actually, the same conclusion holds as soon as the hypotheses are fulfilled for a subsequence $\{n_k\}$ of positive integers instead of the whole sequence of positive integers (see the remarks following Theorem 2.2 in [GS]). It is not known if every hypercyclic operator satisfies the latter general version of the Hypercyclicity Criterion, see [LM1], [LM2], [GLM] and [BP] for more on this subject.

A Beurling type Theorem.

Since the proof of Theorem 2.11 will be accomplished by applying the Hypercyclicity Criterion, we need to know some special dense subsets in $\mathcal{S}_\nu$. The following Lemma is what essentially determines the cut-off of the hypercyclicity of λC_φ and nearly determines the cut-off of the hypercyclicity of C_φ.

LEMMA 2.13. *Let m be any positive integer and α a complex number. If $|\alpha| > 1$, then the subspace of all polynomials that vanish m times at α is dense on any of the $\mathcal{S}_\nu$ spaces. Furthermore, the same is true even if $|\alpha| = 1$, provided that $\nu \leq 1/2$.*

PROOF. To begin with, we prove the result for the subspace $\mathcal{P}$ of all polynomials that vanish at α. Now, a function $f(z) = \sum_{n=0}^\infty a_n z^n \in \mathcal{S}_\nu$ that is orthogonal to $\mathcal{P}$ is, in particular, orthogonal to $z^n - \alpha z^{n-1}$ for each positive integer n. Therefore, as in [Sh2, p. 111], it follows that

$$0 = \langle f, z^n - \alpha z^{n-1} \rangle = (n+1)^{2\nu} a_n - n^{2\nu} \bar{\alpha} a_{n-1}.$$

Upon iterating this identity we see that $a_n = a_0 \bar{\alpha}^n / (n+1)^{2\nu}$. As the series

$$\sum_{n=0}^\infty \frac{|\alpha|^{2n}}{(n+1)^{2\nu}}$$

diverges whenever $|\alpha| > 1$ and any real number ν and, the same is true for $|\alpha| = 1$ provided that $\nu \leq 1/2$, each of the Taylor coefficients in the Taylor expansion of f must be zero. Therefore, f must be zero. As $\mathcal{P}$ is a linear subspace, it must be dense and the statement of the lemma is true for $m = 1$.

Now, suppose that $m > 1$. Let $M_{z-\alpha}$ denote the operator of multiplication by $z - \alpha$ on $\mathcal{S}_\nu$. Since the operator of multiplication by z is bounded on $\mathcal{S}_\nu$, so is $M_{z-\alpha}$. As we already know that $\mathcal{P}$ is dense in $\mathcal{S}_\nu$ (whenever the hypotheses of the Lemma hold), we may conclude that $M_{z-\alpha}$ has dense range in $\mathcal{S}_\nu$. Now, the subspace of polynomials that vanish m times at α is the image under $M_{z-\alpha}^{m-1} = M_{(z-\alpha)^{m-1}}$ of $\mathcal{P}$. Since $M_{z-\alpha}$ has dense range, so does $M_{z-\alpha}^{m-1}$. The result follows because the image of a dense set under an operator with dense range is itself dense. □

The Inner Product Criterion.

The following easy criterion will be very useful to prove the non hypercyclicity of a given operator.

LEMMA 2.14. *Let T be a bounded operator on a separable Hilbert space $\mathcal{H}$. Suppose that for each $f \in \mathcal{H}$ there is a non zero vector $g \in \mathcal{H}$ such that the set of complex numbers*

$$\langle T^n f, g \rangle \tag{11}$$

is not dense in $\mathbb{C}$. Then T is not hypercyclic.

PROOF. Any non zero vector $g \in \mathcal{H}$ defines a linear functional $f \to \langle f, g \rangle$ whose range is $\mathbb{C}$. Suppose that there is $f \in \mathcal{H}$ such that $\{T^n f\}$ is dense in $\mathcal{H}$. Since the image of a dense set under an operator with dense range is itself dense, it follows that the set of complex numbers in (11) is dense, a contradiction. □

We quote here that, with the obvious modifications, Lemma 2.14 also works for Banach spaces or even for Frechet spaces.

In this work, all the applications of Lemma 2.14 will be accomplished by seeing that the set of complex numbers in (11) is bounded. However, in other applications it can be enough to prove that the set in (11) is contained in a straight line or is a constant multiplied by the set of powers of a fixed complex number. This latter argument was used in [CS, Prop. 4.1] and [Bo] to prove a result of Kitai [Ki, Cor. 2.4] that asserts that the adjoint of a hypercyclic operator has always empty point spectrum.

Before proving Theorem 2.11 we still need a couple of technical lemmas. The next one shows that we may ignore the constant functions to prove the non cyclicity of a given composition operator. Let $\mathcal{S}_\nu^0$ be the space that is obtained from $\mathcal{S}_\nu$ by just identifying functions that differ by a constant. The fact that

$$f \circ \varphi - g \circ \varphi$$

is constant whenever f and g differ by a constant shows that C_φ defines a composition operator $\widetilde{C}_\varphi : \mathcal{S}_\nu^0 \to \mathcal{S}_\nu^0$. The following lemma shows that if $\widetilde{C}_\varphi$ is not cyclic, then neither is C_φ.

LEMMA 2.15. *Let φ be an analytic self-map of the unit disk. If C_φ is cyclic on $\mathcal{S}_\nu$, then so is $\widetilde{C}_\varphi$ on $\mathcal{S}_\nu^0$. Furthermore, the same is true if we replace cyclicity by hypercyclicity and C_φ by λC_φ for any $\lambda \neq 0$. In addition, if $\mathrm{span}\{\widetilde{C}_\varphi^n f\}$ has infinite codimension, so does $\mathrm{span}\{C_\varphi^n f\}$.*

PROOF. We only prove the cyclic part. Since the subspace of constant functions is invariant under any composition operator, a constant function cannot be cyclic. On the other hand, for any couple of functions f and g in $\mathcal{S}_\nu$ and any finite sequence $\alpha_1, \ldots, \alpha_n \subset \mathbb{C}$ we have

$$\left\| \sum_{k=1}^n \alpha_k C_{\varphi_k} g - f \right\|_{\mathcal{S}_\nu} \geq \left\| \sum_{k=1}^n \alpha_k C_{\varphi_k} g - f + f(0) - \sum_{k=1}^n \alpha_k g(\varphi_k(0)) \right\|_{\mathcal{S}_\nu}$$
$$= \left\| \sum_{k=1}^n \alpha_k \widetilde{C}_{\varphi_k} g - f \right\|_{\mathcal{S}_\nu^0}$$

from which the result follows. □

Lemma 2.15 is a particular case of a general situation. If T is a linear bounded operator defined on $\mathcal{H} = \mathcal{H}_1 \oplus \mathcal{H}_2$ and $\mathcal{H}_1$ is invariant under T, then the compression of T to $\mathcal{H}_2$ is also hypercyclic (see [He2]). The fact that it is also true for other cyclic properties was noticed in [Sa2]. In what follows, except if there is a risk of confusion, we will denote $\widetilde{C}_\varphi$ by C_φ.

Finally, to prove Theorem 2.11 we need a growth estimate for the first derivative of the $\mathcal{S}_\nu$ functions.

PROPOSITION 2.16 (Growth estimate for the derivative). *If $\nu < 1/2$, then for each $f(z) = \sum_{n=0}^\infty a_n z^n \in \mathcal{S}_\nu$ there is a constant $C > 0$ such that*

$$|f'(z)| \leq \frac{C}{(1-|z|^2)^{(3-2\nu)/2}} \qquad (z \in \mathbb{D}).$$

PROOF. To prove this estimate we first apply the Cauchy-Schwarz inequality to the power series representation of f'

$$|f'(z)| \leq \sum_{n=1}^{\infty} n|a_n||z|^{n-1} \leq \left(\sum_{n=1}^{\infty} |a_n|^2 (n+1)^{2\nu}\right)^{1/2} \left(\sum_{n=1}^{\infty} \frac{|z|^{2n-2}}{(n+1)^{2\nu-2}}\right)^{1/2}.$$

Since $3 - 2\nu > 0$, Stirling's formula shows that

$$\sum_{n=1}^{\infty} (n+1)^{2-2\nu} |z|^{2n-2} \approx \frac{1}{(1-|z|^2)^{3-2\nu}}$$

and the desired growth estimate follows. □

Now, we have all the necessary tools to prove Theorem 2.11.

PROOF OF THEOREM 2.11. First we prove that the conditions are necessary. Since hypercyclicity is invariant under similarity, we may suppose that the boundary fixed point is 1.

First suppose that $\nu > 1/2$. Then the reproducing kernel at 1

$$K_1(z) = \sum_{n=0}^{\infty} \frac{z^n}{(n+1)^{2\nu}}$$

is in $\mathcal{S}_\nu$. In addition, for any $f \in \mathcal{S}_\nu$ we have

$$\langle \bar{\lambda} C_\varphi^\star K_1, f \rangle = \bar{\lambda} \langle K_1, C_\varphi f \rangle = \bar{\lambda} \langle K_1, f \circ \varphi \rangle = \bar{\lambda} f(\varphi(1)) = \bar{\lambda} f(1) = \langle \bar{\lambda} K_1, f \rangle.$$

Thus K_1 is an eigenvector of $\bar{\lambda} C_\varphi^\star$. Since the adjoint of a hypercyclic operator has empty point spectrum, it follows that λC_φ cannot be hypercyclic for any λ (see [Ki, Cor. 2.4]).

Now, suppose that $\nu = 1/2$, that is, $\mathcal{S}_\nu$ is the Dirichlet space $\mathcal{D}$. In this case the argument above does not work but the non hypercyclicity of C_φ is rather trivial and a consequence of a very basic property, that is, the norm of a hypercyclic operator must be strictly greater than one. Indeed, if λC_φ is hypercyclic on the Dirichlet space for $|\lambda| \leq 1$, then, by Lemma 2.15, so is on $\mathcal{D}_0$. But on $\mathcal{D}_0$ we have

$$\int_{\mathbb{D}} |\lambda f'(\varphi(z))|^2 |\varphi'(z)|^2 \, dA(z) = |\lambda|^2 \int_{\varphi(\mathbb{D})} |f'(z)|^2 \, dA(z) \leq \int_{\mathbb{D}} |f'(z)|^2 \, dA(z).$$

Thus $\|\lambda C_\varphi\|_{\mathcal{D}_0} \leq 1$, and therefore, C_φ cannot be hypercyclic on $\mathcal{D}_0$, a contradiction. Observe that we have also proved half part of (ii).

It remains to prove that the conditions are necessary for $\nu < 1/2$. Now, we need an easy expression for φ. We have already assumed that 1 is the boundary fixed point. First, we perform the change of variables

$$\sigma(z) = \frac{i(1+z)}{1-z}$$

that sends the unit disk onto the upper half plane and 1 to ∞ and the exterior fixed point to a point p in the lower half plane. Upon conjugating with an appropriate affine map in the upper half plane we may suppose that p is on the imaginary axis.

Finally, coming back to the unit disk, σ^{-1} sends p onto a negative number $\alpha < -1$. Therefore, we may suppose that φ has the expression

$$\varphi(z) = \frac{(\mu\alpha - 1)z + \alpha(1 - \mu)}{(\mu - 1)z + \alpha - \mu} \qquad \text{with } 0 < \mu < 1. \qquad (12)$$

Finally, we conjugate one more time with

$$\frac{\alpha z - 1}{\alpha - z}$$

that is an automorphism of the unit disk and fixes 1 and sends α to ∞ and, therefore, we may suppose that

$$\varphi(z) = \mu z + 1 - \mu \qquad \text{with } 0 < \mu < 1. \qquad (13)$$

Observe that $\varphi'(1) = \mu$. The use of formula (12) or (13) will depend on the computations that we have to do. Formula (13) above allows us to obtain the following easy expression for the iterates

$$\varphi_n(z) = \mu^n z + 1 - \mu^n. \qquad (14)$$

Now, let λ be a complex number with $|\lambda| \leq \mu^{(1-2\nu)/2}$ and consider any $f \in \mathcal{S}_\nu$. The growth estimate for the derivative provides a constant C such that the first inequality below holds. The condition on λ is applied in the second inequality below

$$\begin{aligned}
|\langle \lambda^n C_{\varphi_n} f, z \rangle| &= 2^{2\nu} |\lambda|^n |(f \circ \varphi_n)'(0)| \\
&= 2^{2\nu} |\lambda|^n |f'(\varphi_n(0))| \, |\varphi_n'(0)| \\
&= 2^{2\nu} |\lambda|^n |f'(1 - \mu^n)| \mu^n \\
&\leq \frac{2^{2\nu} C |\lambda|^n \mu^n}{(1 - |1 - \mu^n|^2)^{(3-2\nu)/2}} \\
&\leq \frac{2^{2\nu} C \mu^{n(3-2\nu)/2}}{(2\mu^n - \mu^{2n})^{(3-2\nu)/2}} \\
&= \frac{2^{2\nu} C}{2 - \mu}.
\end{aligned}$$

Upon applying Lemma 2.14 we see that λC_φ is not hypercyclic. Therefore, we have completed the proof that the conditions are necessary.

Now we prove that the conditions are also sufficient, that is, we suppose that $\nu \leq 1/2$ and $|\lambda| > \mu^{(1-2\nu)/2}$. Let X be the set of all polynomials that vanish m times at 1, where m will be determined in the course of the proof. Due to Lemma 2.13 the set X is dense in $\mathcal{S}_\nu$.

Let $p \in X$ be fixed. For $\nu < 0$ we can use the integral representation of the norm given in (7) (in fact, it is easier), but to handle all cases simultaneously, we will use the equivalent norm furnished by Lemma 1.2 with $l = 0$. We set $c = 1 - 2\nu$. Upon performing the change of variables $w = \varphi_n(z)$ in the second equality below

and the change of variables $w = \varphi_{-n}(z)$ in the third equality below we have

$$\begin{aligned}
\|C_{\varphi_n} p\|^2 &= |p(\varphi_n(0))|^2 + \int_{\mathbb{D}} |p'(\varphi_n(z))|^2 |\varphi_n'(z)|^2 (1-|z|^2)^c dA(z) \\
&= |p(\varphi_n(0))|^2 + \int_{\varphi_n(\mathbb{D})} |p'(z)|^2 (1-|\varphi_{-n}(z)|^2)^c dA(z) \\
&\leq |p(\varphi_n(0))|^2 + \max_{\varphi_n(\mathbb{D})} |p'(z)|^2 \int_{\varphi_n(\mathbb{D})} (1-|\varphi_{-n}(z)|^2)^c dA(z) \\
&= |p(\varphi_n(0))|^2 + \mu^{2n} \max_{\varphi_n(\mathbb{D})} |p'(z)|^2 \int_{\mathbb{D}} (1-|z|^2)^c dA(z) \\
&= |p(\varphi_n(0))|^2 + C\mu^{2n} \max_{\varphi_n(\mathbb{D})} |p'(z)|^2,
\end{aligned}$$

where $C = \|z\|^2$. Now, $p(z) = (1-z)^m q(z)$ where q is a polynomial. By denoting by C_1 the maximum of $|q(z)|^2$ on $\overline{\mathbb{D}}$ we have

$$|p(\varphi_n(0))|^2 = |p(1-\mu^n)|^2 = |\mu^{mn}|^2 |q(1-\mu^n)|^2 \leq C_1 \mu^{2mn}.$$

On the other hand, $p'(z) = (z-1)^{m-1} q_1(z)$ where $q_1(z)$ is another polynomial. Let C_2 be the maximum of $|q_1(z)|^2$ on $\overline{\mathbb{D}}$. Now, observe that $\varphi_n(\mathbb{D})$ is a disk of radius $(1-2\mu^n)/2$ which is interiorly tangent to the unit disk at 1. The maximum of $|1-z|^{2m-2}$ on $\varphi_n(\mathbb{D})$ is attained at $\varphi_n(-1) = 1-2\mu^n$ and its value is $2^{2m-2}\mu^{2(m-1)n}$. Therefore, we have

$$\max_{\varphi_n(\mathbb{D})} |p'(z)|^2 \leq 2^{2m-2} C_2 \mu^{2(m-1)n}.$$

Thus putting everything together we find that

$$\|\lambda^n C_\varphi^n p\| \leq M |\lambda \mu^m|^n,$$

where M is a constant which is independent of n. Since $0 < \mu < 1$, we can choose m large enough to have $|\lambda \mu^m| < 1$ (just take $m > -\log|\lambda|/\log\mu$). Thus, for $\nu < 0$ the iterates $\lambda^n C_{\varphi_n}$ tend to zero pointwise on X as $n \to \infty$.

To verify hypothesis (b) of the Hypercyclicity Criterion it is easier to use formula (2) instead of (1). We are free to do so because both formulae induce similar composition operators. Thus the iterates of $\varphi_{-1} = \varphi^{-1}$ are

$$\varphi_{-n}(z) = \frac{(\mu^{-n}\alpha - 1)z + \alpha(1-\mu^{-n})}{(\mu^{-n}-1)z + \alpha - \mu^{-n}} \quad \text{with } 0 < \mu < 1 \text{ and } \alpha < -1, \quad (15)$$

where n ranges through all the non negative integers.

The inverse map S will be $\lambda^{-1} C_{\varphi_{-1}}$ that assigns to each function f the function $\lambda^{-1} f \circ \varphi_{-1}$. This map is not bounded on $\mathcal{S}_\nu$ but it is still well defined on the polynomials and boundedness of S is not required by the Hypercyclicity Criterion. Let Y_0 be the set of all polynomials that vanish m times at α where m will be determined later on. The set Y required by the Hypercyclicity Criterion will be

$$Y = \bigcup_{n=0}^\infty \lambda^{-n} C_{\varphi_{-1}}^n (Y_0) = \bigcup_{n=0}^\infty \lambda^{-n} C_{\varphi_{-n}}(Y_0).$$

Obviously, $\lambda^{-1} C_{\varphi_{-1}}$ takes Y into itself and it is a right inverse of λC_φ on Y. Observe that if we show that $S^n = \lambda^{-n} C_{\varphi_{-n}}$ tends pointwise to zero on Y_0, then so does on Y.

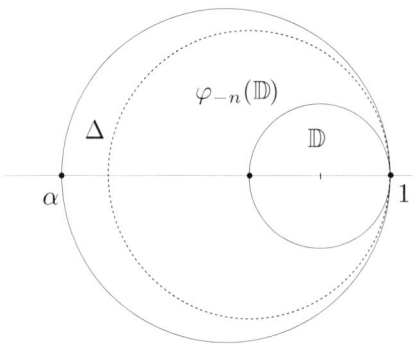

Figure 1

Now, suppose that $p \in Y_0$ is fixed. We will estimate
$$\|C_{\varphi_{-n}}p\|^2 = \|p \circ \varphi_{-n}\|^2.$$

Let Δ denote the disk that touches tangentially the unit disk at 1 and passes through the exterior fixed point α. Observe that $\varphi_{-n}(\mathbb{D})$ is contained in Δ for every n. Indeed, $\varphi_{-n}(\mathbb{D})$ approaches Δ as n tends to ∞ (see Figure 1). Now, $p(z) = (z-\alpha)^m q(z)$ where q is a polynomial. We set $C_1 = \max_{\overline{\Delta}} |q(z)|^2$. Thus we have
$$|p(\varphi_{-n}(0))|^2 \leq C_1 |(\varphi_{-n}(0) - \alpha)^m|^2.$$

The change of variables $w = \varphi_{-n}(z)$ shows that
$$\int_{\mathbb{D}} |p'(\varphi_{-n}(z))|^2 |\varphi'_{-n}(z)|^2 (1-|z|^2)^c \, dA(z) = \int_{\varphi_{-n}(\mathbb{D})} |p'(z)|^2 (1-|\varphi_n(z)|^2)^c \, dA(z).$$

Since $p'(z) = (z-\alpha)^{m-1} q_1(z)$ where q_1 is another polynomial, the integral above is less than
$$C_2 \int_{\varphi_{-n}(\mathbb{D})} |(z-\alpha)^{m-1}|^2 (1-|\varphi_n(z)|^2)^c \, dA(z),$$
where $C_2 = \max_{\overline{\Delta}} |q_1(z)|^2$. Now, the change of variables $w = \varphi_n(z)$ shows that the integral above is equal to
$$C_2 \int_{\mathbb{D}} |(\varphi_{-n}(z)-\alpha)^{m-1}|^2 |\varphi'_{-n}(z)|^2 (1-|z|^2)^c \, dA(z).$$

Upon putting everything together we find that
$$\|p \circ \varphi_{-n}\|^2 \leq C \|(\varphi_{-n}(z) - \alpha)^m\|^2, \tag{16}$$

where C is a constant independent of n. From now on, we will use the definition of the norm that is given by the Taylor coefficients. This is not only for reasons of variety but also because we can give a proof for all ν simultaneously (if we used the integral representation, then we would have to distinguish two cases $c > 0$ and $0 \geq c > -1$).

In the remainder of the proof C will denote a constant that may be different from one display to another but it is always independent of n. Upon substituting (15) in (16) we have

$$\|C_{\varphi_{-n}} p\|^2 \leq C \left\| \frac{(\alpha-1)^m (z-\alpha)^m}{((\mu^{-n}-1)z+\alpha-\mu^{-n})^m} \right\|^2. \tag{17}$$

Since the operator of multiplication by $z - \alpha$ is bounded on any of the $\mathcal{S}_\nu$ spaces, so is the operator of multiplication by $(z-\alpha)^m$. Hence, we find that (17) is less than or equal to

$$C \left\| \frac{1}{((\mu^{-n}-1)z+\alpha-\mu^{-n})^m} \right\|^2. \tag{18}$$

Now, since

$$\frac{1}{((\mu^{-n}-1)z+\alpha-\mu^{-n})^m} = \frac{\mu^{mn}}{(\alpha\mu^n-1)^m} \sum_{k=0}^{\infty} \frac{\Gamma(k+m)}{\Gamma(k+1)\Gamma(m)} \left(\frac{\mu^n-1}{\alpha\mu^n-1}\right)^k z^k,$$

we have, by the definition of the norm on $\mathcal{S}_\nu$, that (18) is equal to

$$C \frac{\mu^{2mn}}{|1-\alpha\mu^n|^{2m}} \sum_{k=0}^{\infty} \frac{\Gamma^2(k+m)}{\Gamma^2(k+1)\Gamma^2(m)} (k+1)^{2\nu} \left|\frac{\mu^n-1}{1-\alpha\mu^n}\right|^{2k}. \tag{19}$$

Stirling's formula shows that (19) is less than

$$C \frac{\mu^{2mn}}{|1-\alpha\mu^n|^{2m}} \sum_{k=0}^{\infty} k^{2m+2\nu-2} \left|\frac{\mu^n-1}{1-\alpha\mu^n}\right|^{2k}. \tag{20}$$

On the other hand, for $a > 0$ Stirling's formula shows that

$$\sum_{k=1}^{\infty} k^{a-1} |z|^{2k} \approx \frac{1}{(1-|z|^2)^a}.$$

Therefore, if we choose m large enough to have $2m + 2\nu - 1 > 0$, we find that (20) is less than

$$\frac{C\mu^{2mn}}{|1-\alpha\mu^n|^{2m}} \left(1 - \left|\frac{1-\mu^n}{1-\alpha\mu^n}\right|^2\right)^{1-2m-2\nu} = \frac{C|1-\alpha\mu^n|^{2m+4\nu-2} \mu^{n(1-2\nu)}}{((\alpha^2-1)\mu^n+2(1-\alpha))^{2m+2\nu-1}}.$$

Thus having in mind that $0 < \mu < 1$ and $\alpha < -1$ we have for n large enough

$$\|\lambda^{-n} C_{\varphi_{-n}} p\| \leq C |\lambda|^{-n} \mu^{n(1-2\nu)/2},$$

where C is a constant that depends on α, m and ν but not on n. Finally, since $|\lambda|^{-1} \mu^{(1-2\nu)/2} < 1$, it follows that the iterates of $\lambda^{-1} C_{\varphi_{-1}}$ tend to zero pointwise on Y_0 and, consequently, so do on Y. Therefore, all the hypotheses of the Hypercyclicity Criterion are fulfilled. Thus the conditions are also sufficient. □

Invariant subspaces of finite codimension.

If $\nu > 1/2$ and $|\alpha| = 1$ the statement of Lemma 2.13 is no longer true. In fact, let k be the unique positive integer such that $(2k-1)/2 < \nu \le (2k+1)/2$ and consider the functions

$$K_w^{(j)}(z) = \sum_{n=j}^{\infty} \frac{n!}{(n-j)!} \frac{\bar{w}^{n-j}}{(n+1)^{2\nu}} z^n, \qquad j = 0, 1, \ldots, k-1,$$

which are the reproducing kernel at w and its successive derivatives with respect to w (see [CM] for more details). These functions belong to $\mathcal{S}_\nu$ even if $w = \alpha \in \partial \mathbb{D}$ because $\nu > (2k-1)/2$. One easily checks that

$$f^{(j)}(\alpha) = \langle f, K_\alpha^{(j)} \rangle \quad \text{for } j = 0, 1, \ldots, k-1. \tag{21}$$

Therefore, if p is any polynomial that vanishes, at least, k times at α, then it is orthogonal to $K_\alpha^{(j)}$, $0 \le j \le k-1$, and Lemma 2.13 fails to be true. However, each of the expressions in (21) defines a linear functional on $\mathcal{S}_\nu$ whose respective kernel is precisely the orthogonal complement to $K_\alpha^{(j)}$. Thus the set of polynomials that vanish, at least, k times at α is dense in the orthogonal complement of span $\{K_\alpha^{(j)} : 0 \le j \le k-1\}$. This latter space will be denoted by $\mathcal{S}_{\nu,k}$. Now, the fact that $\mathcal{S}_{\nu,k}$ is invariant under the multiplication operator $M_{z-\alpha}$ allows us to repeat the last argument of the proof of Lemma 2.13 to obtain the following

LEMMA 2.17. *Suppose that $\nu > 1/2$ and $|\alpha| = 1$. Let k be the unique integer such that $(2k-1) < 2\nu \le (2k+1)$. Then given any integer $m \ge k$, the set of all polynomials that vanish m times at α is dense in $\mathcal{S}_{\nu,k}$.*

With Lemma 2.17 at hand we can prove the following Theorem that gives hypercyclicity of scalar multiples on an invariant subspace of codimension k.

THEOREM 2.18. *Let φ be a hyperbolic non-automorphism and consider $C_\varphi : \mathcal{S}_\nu \to \mathcal{S}_\nu$ where $\nu > 1/2$. Then there is a complex number λ such that λC_φ is hypercyclic on a C_φ-invariant subspace of codimension k, where k is the first integer such that $2k + 1 \ge 2\nu$.*

Recall that for $\nu > 1/2$ the space $\mathcal{S}_\nu$ consists of analytic functions that extend continuously to the boundary of $\mathbb{D}$. A Hilbert space of analytic functions on $\mathbb{D}$ is said to be a natural space if norm convergence implies uniform convergence on compact subsets of $\mathbb{D}$. On these spaces no continuous function on $\overline{\mathbb{D}}$ can be hypercyclic for a bounded composition operator because every function would be bounded on any compact subset of $\mathbb{D}$. However, Theorem 2.18 has the following intriguing corollary.

COROLLARY 2.19. *There exists a natural space $\mathcal{H}$ consisting of analytic functions on $\mathbb{D}$ and continuous on $\overline{\mathbb{D}}$ and a bounded composition operator C_φ on $\mathcal{H}$ such that for some complex number λ the operator λC_φ is hypercyclic on $\mathcal{H}$.*

PROOF OF THEOREM 2.18. Again, we suppose that 1 is the boundary fixed point of φ. We already know that for each $0 \le j \le k-1$ the function $K_1^{(j)}(z) \in \mathcal{S}_\nu$. In addition,

$$\langle C_\varphi^\star K_1^{(j)}, f \rangle = \langle K_1^{(j)}, C_\varphi f \rangle = (f \circ \varphi)^{(j)}(1) = \mu^j f^{(j)}(1) = \mu^j \langle K_1^{(j)}, f \rangle.$$

Thus $K_1^{(j)}$ is an eigenvector for the adjoint $C_\varphi^\star$. Consequently, it follows that if f is orthogonal to each $K_1^{(j)}$, then so is $C_\varphi f$ and we may conclude that $\mathcal{S}_{\nu,k}$ is invariant under C_φ.

We will prove that if $|\lambda| > \mu^{(1-2\nu)/2}$, then λC_φ is hypercyclic on $\mathcal{S}_{\nu,k}$. We do not verify hypothesis (b) of the Hypercyclicity Criterion because it follows in a similar way to that of the proof of Theorem 2.11.

The set X will be the set of all polynomials that vanish m times at 1, where $m > k$ is to be determined later on. By Lemma 2.17, X is dense in $\mathcal{S}_{\nu,k}$.

Let $p \in X$ be fixed. We will use the norm furnished by Lemma 1.2 with $l = k$. We set $c = 2k + 1 - 2\nu$. By the expression for the iterates in (14), we have $(p \circ \varphi_n)^{(j)}(z) = \mu^{nj} p^{(j)}(\varphi_n(z))$ for each pair of positive integers j and n. Therefore,

$$\|C_\varphi^n p\|^2 = \sum_{j=0}^{k} \mu^{2jn} |p^{(j)}(1-\mu^n)|^2 + \mu^{2n(k+1)} \int_{\mathbb{D}} |p^{(k+1)}(\varphi_n(z))|^2 (1-|z|^2)^c \, dA(z). \tag{22}$$

As p vanishes m times at 1, for $0 \leq j \leq k+1$ we can write $p^{(j)}(z) = (1-z)^{m-j} q_j(z)$ where each $q_j(z)$, $0 \leq j \leq k+1$, is a polynomial. Let C be a common bound for $|q_j(z)|^2$ in $\overline{\mathbb{D}}$ for $0 \leq j \leq k+1$. Therefore, the first term in the right hand side of (22) is equal to

$$\sum_{j=0}^{k} \mu^{2jn} \mu^{2n(m-j)} |q_j(1-\mu^n)|^2 \leq (k+1) C \mu^{2nm}.$$

As before, we perform the change of variables $w = \varphi_n(z)$ in the integral in (22). Then, we take $\max_{\overline{\varphi_n(\mathbb{D})}} |p^{(k+1)}(z)|^2$ out of the integral. Finally, undoing the change of variables we find that the integral in (22) is less than

$$\mu^{2n(k+1)} \max_{\overline{\varphi_n(\mathbb{D})}} |p^{(k+1)}(z)|^2 \int_{\mathbb{D}} (1-|z|^2)^c \, dA(z). \tag{23}$$

Now, it is easy to check that

$$\max_{\overline{\varphi_n(\mathbb{D})}} |(1-z)|^{2m-2k-2} = 2^{2m-2k-2} \mu^{2n(m-k-1)}.$$

Therefore, we find that (23) is less than

$$C 2^{2m-2k-2} \mu^{2mn}.$$

Hence, there is a constant M independent of n such that

$$\|\lambda^n C_\varphi^n p\| \leq M |\lambda \mu^m|^n.$$

Thus, for $m > k$ large enough, we have $|\lambda \mu^m|$ is less than 1. So we can conclude that (14) tends to zero. Therefore, the hypotheses of the Hypercyclicity Criterion are fulfilled and the proof Theorem 2.18 is concluded. $\square$

REMARK 2.20. To prove that the iterates of λC_φ goes to zero on X no hypothesis on λ has been used. However, the greater $|\lambda|$ is, the greater the integer m we had to choose is. On the other hand, for $|\lambda|$ large enough the iterates of $\lambda^{-1} C_\varphi^{-1}$ always goes to zero on the polynomials that vanish m times at α; here m only depends on ν. Thus we can say that the cut-off of the hypercyclicity of λC_φ is really determined by the density of the set X, that is, Lemma 2.13.

REMARK 2.21. The spectrum of C_φ acting on $\mathcal{S}_\nu$ with $\nu \leq 1/2$ is the closed unit disk centered at the origin of radius $\varphi'(\eta)^{(2\nu-1)/2}$ (see [Hu, Thm. 8 and Cor. 12]). Therefore, we can restate part (i) of Theorem 2.11 by saying that λC_φ is hypercyclic if and only if $1/\lambda$ is in the interior of the spectrum of C_φ. In fact, we could have used a result of Kitai that asserts that the spectrum of a hypercyclic operator must meet the unit circle [Ki, Thm. 2.8] to see that if $|\lambda| < \varphi'(\eta)^{(1-2\nu)/2}$, then λC_φ cannot be hypercyclic. However, Kitai's result does not work for $|\lambda| = \varphi'(\eta)^{(2\nu-1)/2}$.

REMARK 2.22. Theorem 2.18 is the best possible in the following sense: On one hand, for any complex number λ the operator λC_φ cannot be hypercyclic on $\mathcal{S}_{\nu,k-1}$ because $K_1^{(k-1)}$ is an eigenvector for the adjoint $\bar{\lambda} C_\varphi^\star$. On the other hand, C_φ itself cannot be hypercyclic on $\mathcal{S}_{\nu,k}$. If $\nu > 1/2$ the spectrum of C_φ acting on $\mathcal{S}_\nu$ is (see [Hu, Cor. 12])

$$\{t : |t| \leq \varphi'(\eta)^{(2\nu-1)/2}\} \cup \{\varphi'(\eta)^j : j = 0, 1, \dots\}.$$

Thus we may deduce that the spectrum of C_φ acting on $\mathcal{S}_{\nu,k}$ is the closed disk $\overline{D}(0,R)$ with $R = \varphi'(\eta)^{(2\nu-1)/2} < 1$. Moreover, using Lemma 2.14 it is easy to see that for $|\lambda| \leq \varphi'(\eta)^{(1-2\nu)/2}$ the operator λC_φ is not hypercyclic on $\mathcal{S}_{\nu,k}$. Of course, by filling up the details as in the proof of Theorem 2.11 one easily checks that λC_φ is hypercyclic on $\mathcal{S}_{\nu,k}$ if and only if $|\lambda|$ is strictly greater than the inverse of the spectral radius of C_φ acting on $\mathcal{S}_{\nu,k}$.

CHAPTER 3

NON ELLIPTIC AUTOMORPHISMS

In this chapter we deal with the case in which φ is a hyperbolic or a parabolic automorphism. Further along this chapter we will prove the following

THEOREM 3.1. *Let φ be a non-elliptic automorphism of the unit disk. Then*
- (a) *If φ is a parabolic automorphism, then λC_φ is hypercyclic on $\mathcal{S}_\nu$ if and only if $\nu < 1/2$ and $|\lambda| = 1$.*
- (b) *If φ is a hyperbolic automorphism and η is its attractive fixed point, then λC_φ is hypercyclic on $\mathcal{S}_\nu$ if and only if $\nu < 1/2$ and $\varphi'(\eta)^{(1-2\nu)/2} < |\lambda| < \varphi'(\eta)^{(2\nu-1)/2}$.*
- (c) *The operator C_φ is hypercyclic if and only if $\nu < 1/2$.*
- (d) *The operator C_φ is cyclic if and only if $\nu < 1/2$.*

Part (d) of Theorem 3.1 gives exactly the cut-off of the cyclicity of C_φ and answers a question posed by Zorboska [Zo3]. This latter author proved that if φ is a hyperbolic automorphism then C_φ is not cyclic for $\nu > 1/2$ and that if φ is a parabolic automorphism, then C_φ is not cyclic for $\nu > 3/2$. In Chapter 4, we will see that the cut-off of the cyclicity in the case of the parabolic non automorphism is different from that of the parabolic automorphism. Observe that composition operators induced by non elliptic automorphisms are cyclic if and only if they are hypercyclic. Actually, there is an oversight in the statement of Proposition 3.5 in [Zo3] in which it is asserted that C_φ on the Dirichlet space ($\nu = 1/2$) is hypercyclic.

HYPERCYCLICITY

First we will dispense with the hypercyclicity of λC_φ. Although there are still unbounded functions in the Dirichlet space, these functions cannot be hypercyclic vectors. The point is that every function $f \in \mathcal{D}$ satisfies the following growth estimate
$$|f(z)| \leq C \log(1 - |z|),$$
that prevents functions in the Dirichlet space to be hypercyclic. In fact, the cut-off for non elliptic automorphisms at the Dirichlet space is very drastic. This is in a strong contrast with the case in which φ is a hyperbolic non automorphism (Chapter 2) or a parabolic non automorphism (Chapter 4). Moreover, Theorems 3.13 and 3.14 will show that C_φ acting on the Dirichlet space is strongly non cyclic.

We will need the following easy proposition about composition operators induced by automorphisms acting on $\mathcal{D}_0$, the Dirichlet space modulo constants.

PROPOSITION 3.2. *Let φ be an automorphism of $\mathbb{D}$. Then C_φ acting on $\mathcal{D}_0$ is a unitary isometry.*

PROOF. First, for all $f, g \in \mathcal{D}_0$ we have

$$\langle C_\varphi f, g \rangle = \int_\mathbb{D} f'(\varphi(z)) \varphi'(z) \overline{g'(z)} \, dA(z)$$

The change of variables $z = \varphi_{-1}(w) = \varphi^{-1}(w)$ yields $dA(z) = |\varphi'_{-1}(z)|^2 dA(w)$ and $\varphi'(z) = 1/\varphi'_{-1}(w)$. Thus the integral above is equal to

$$\int_\mathbb{D} f(w) \overline{g(\varphi_{-1}(w)) \varphi'_{-1}(w)} \, dA(w) = \langle f, C_{\varphi_{-1}} g \rangle.$$

Therefore, $C_\varphi^\star = C_{\varphi_{-1}} = C_\varphi^{-1}$. Consequently, C_φ is a unitary isometry. □

The parabolic automorphism.

THEOREM 3.3. *Let φ be a parabolic automorphism of the unit disk. Then λC_φ is hypercyclic on $\mathcal{S}_\nu$ if and only if $\nu < 1/2$ and $|\lambda| = 1$.*

PROOF. First we prove that the conditions are necessary. Let us start by examining the case $\nu = 1/2$. Suppose now that $|\lambda| \leq 1$. Since C_φ acting on $\mathcal{D}_0$ is a unitary isometry, we find that $\|\lambda C_\varphi\|_{\mathcal{D}_0} \leq 1$ and, consequently, λC_φ cannot be hypercyclic on $\mathcal{D}_0$. By applying Lemma 2.15, we see that λC_φ is not hypercyclic on $\mathcal{D}$. If $|\lambda| > 1$, then, by what we have just proved, $\lambda^{-1} C_{\varphi_{-1}}$ is not hypercyclic. Now, an invertible operator is hypercyclic if and only if its inverse is (see [Ki, Cor. 2.2] or [HK]). Consequently, λC_φ is not hypercyclic on $\mathcal{D}$.

For $\nu > 1/2$, the Comparison Principle shows that λC_φ is not hypercyclic on $\mathcal{S}_\nu$ for any λ. Alternatively, we could have used the fact that the reproducing kernel at 1 in $\mathcal{S}_\nu$ is an eigenvector for the adjoint $\bar{\lambda} C_\varphi^\star$.

Now suppose that $\nu < 1/2$. We need an expression for the iterates of φ. By Proposition 1.4 the fixed point of φ must be on the boundary of the unit disk. As usual we may suppose that the fixed point is 1. The change of variables

$$\sigma(z) = \frac{i(1+z)}{1-z}$$

takes the unit disk onto the upper half plane. Thus φ satisfies the following formula

$$\varphi(z) = \frac{(2-a)z + a}{-az + 2 + a} \qquad \text{with } a \neq 0 \text{ and } \Re a = 0. \tag{1}$$

The fact that $\Re a = 0$ reflects that φ corresponds to an automorphism of the upper half plane. It follows that the formula for the iterates is

$$\varphi_n(z) = \frac{(2-na)z + na}{-naz + 2 + na}, \tag{2}$$

which is also valid for n ranging through the whole set of integers.

Now suppose that $|\lambda| < 1$. By the growth estimate for the derivative, Proposition 2.16, there is a constant C such that the inequality below is true

$$\begin{aligned}
|\langle \lambda^n C_{\varphi_n} f, z \rangle| &= 2^{2\nu} |\lambda|^n |f'(\varphi_n(0))| |\varphi_n'(0)| \\
&= 2^{2\nu} |\lambda|^n \left| f'\left(\frac{na}{2+na}\right) \right| \frac{4}{4+n^2|a|^2} \\
&\leq 2^{2\nu+2} C |\lambda|^n \left(1 - \frac{n^2|a|^2}{4+n^2|a|^2}\right)^{(2\nu-3)/2} \frac{1}{4+n^2|a|^2} \\
&= \frac{2^{4\nu-1} C |\lambda|^n}{(4+n^2|a|^2)^{(2\nu-1)/2}}.
\end{aligned}$$

Since the last quantity remains bounded as n goes to ∞, we can apply Lemma 2.14 to see that λC_φ is not hypercyclic. If $|\lambda| > 1$, then $\lambda^{-1} C_{\varphi_{-1}}$ is not hypercyclic and, therefore, neither is λC_φ. Thus the conditions are necessary.

Now, we prove the converse. Thus suppose that $\nu < 1/2$ and $|\lambda| = 1$. Since for $|\lambda| = 1$ an operator T satisfies the Hypercyclicity Criterion if and only if λT does, we need only prove that C_φ satisfies the Criterion. We will use the equivalent norm given by Lemma 1.2. We set $c = 1 - 2\nu$. By the Comparison Principle it is enough to consider the case in which $0 < \nu < 1/2$. Thus we may suppose that $0 < c < 1$.

Let X be the set of all analytic functions that extend continuously to the boundary and vanish at 1. Since X contains the polynomials that vanish at 1, it is dense in $\mathcal{S}_\nu$. Let $f \in X$ be fixed. We have

$$\|C_{\varphi_n} f\|^2 = |f(\varphi_n(0))|^2 + \int_{\mathbb{D}} |f'(\varphi_n(z))|^2 |\varphi_n'(z)|^2 (1-|z|^2)^c \, dA(z). \tag{3}$$

As $\varphi_n(0) \to 1$, we have $|f(\varphi_n(0))|^2 \to |f(1)|^2 = 0$. Hence, we only have to show that the integral in (3) goes to zero as $n \to \infty$. The change of variables $w = \varphi_n(z)$ shows that the integral in (3) is equal to

$$\int_{\mathbb{D}} |f'(z)|^2 (1-|\varphi_{-n}(z)|^2)^c \, dA(z) = \frac{4^c}{|na|^{2c}} \int_{\mathbb{D}} \frac{|f'(z)|^2 (1-|z|^2)^c}{|-1+2/(na)+z|^{2c}} \, dA(z), \tag{4}$$

where in the equality above we have used the expression for φ_{-n} in (2) and the identity $a + \bar{a} = 0$. Now for $z \in \mathbb{D}$ the reverse triangle inequality and the fact that $c > 0$ imply that

$$|-1+2/(na)+z|^{2c} \geq (|1-2/(na)| - |z|)^{2c} > (1-|z|)^{2c}.$$

Therefore, (4) is less than or equal to

$$\frac{2^{2c}}{|na|^{2c}} \int_{\mathbb{D}} |f'(z)|^2 \left(\frac{1+|z|}{1-|z|}\right)^c \leq \frac{2^{3c}}{|na|^{2c}} \max_{\overline{\mathbb{D}}} |f'(z)|^2 \int_{\mathbb{D}} (1-|z|)^{-c} \, dA(z). \tag{5}$$

Since $c < 1$, the singularity is integrable and thus (5) goes to zero because $c > 0$.

For right inverse we take $C_\varphi^{-1} = C_{\varphi_{-1}}$. If we take $Y = X$, then $C_{\varphi_{-1}}$ maps Y into itself and similar computations show that $C_{\varphi_{-1}}^n = C_{\varphi_{-n}}$ also tends pointwise on Y. By applying the Hypercyclicity Criterion we see that C_φ is hypercyclic, which is the desired result. $\square$

REMARK 3.4. For $\nu \leq 0$ the spectrum of a composition operator induced by a parabolic automorphism is the unit circle (see [No], [Cw2] and [Cw3]). Thus, when φ is a parabolic automorphism λC_φ is hypercyclic if and only if λ belongs to the spectrum. As far as we know, for $0 < \nu < 1/2$ the spectrum of C_φ is unknown. Thus, in this case, we cannot apply the result of Kitai that asserts that the spectrum of a hypercyclic operator must meet the unit circle [Ki, Cor. 2.4]. Theorem 3.3 suggests that the spectrum of C_φ should maintain the same formula for $0 < \nu < 1/2$ as for $\nu \leq 0$.

The hyperbolic automorphism.

THEOREM 3.5. *Let φ be a hyperbolic automorphism of the unit disk and η its attractive fixed point. Then λC_φ is hypercyclic if and only if $\nu < 1/2$ and $\varphi'(\eta)^{(1-2\nu)/2} < |\lambda| < \varphi'(\eta)^{(2\nu-1)/2}$.*

PROOF. For $\nu \geq 1/2$, the non hypercyclicity of λC_φ follows exactly as in the case of the parabolic automorphism.

Now, suppose that $\nu < 1/2$. Let us obtain an expression for the iterates of φ. Without loss of generality we may suppose that φ has -1 and 1 as its fixed points. Moreover, we may assume that 1 is the attractive fixed point. We compute φ explicitly by employing again the change of variables

$$\sigma(z) = \frac{i(1-z)}{1+z}$$

that sends the unit disk onto the upper half plane, the fixed points 1 and -1 to 0 and ∞, respectively, and φ to the contraction map $\varphi(w) = \mu w$ where $0 < \mu < 1$. Coming back to the unit disk we have

$$\varphi(z) = \frac{(1+\mu)z + 1 - \mu}{(1-\mu)z + 1 + \mu} \quad \text{with} \quad 0 < \mu < 1,$$

from which we can easily obtain the following formula for the iterates

$$\varphi_n(z) = \frac{(1+\mu^n)z + 1 - \mu^n}{(1-\mu^n)z + 1 + \mu^n}, \tag{6}$$

which is valid for n ranging through the whole set of integers. Observe that the derivative at the attractive fixed point is $\varphi'(1) = \mu$.

Now, for any $f \in \mathcal{S}_\nu$ we have the following estimate for some constant C independent of n

$$\begin{aligned}|\langle \lambda^n C_{\varphi_n} f, z\rangle| &= 2^{2\nu}|\lambda|^n |f'(\varphi_n(0))| \, |\varphi'_n(0)| \\ &= 2^{2\nu}|\lambda|^n \left| f'\left(\frac{1-\mu^n}{1+\mu^n}\right)\right| \frac{4\mu^n}{(1+\mu^n)^2} \\ &\leq 2^{2\nu} C |\lambda|^n \left(1 - \left|\frac{1-\mu^n}{1+\mu^n}\right|^2\right)^{(2\nu-3)/2} \frac{4\mu^n}{(1+\mu^n)^2} \\ &= \frac{2^{4\nu-1} C |\lambda|^n \mu^{n(2\nu-1)/2}}{(1+\mu^n)^{2\nu-1}},\end{aligned}$$

that remains bounded for $|\lambda|\mu^{(2\nu-1)/2} \leq 1$. Therefore, if λC_φ is hypercyclic, then $|\lambda| > \mu^{(1-2\nu)/2}$. In addition, the inverse operator $\lambda^{-1} C_{\varphi_{-1}}$ must also be hypercyclic. The attractive fixed point of φ_{-1} is -1 and $\varphi'_{-1}(-1) = \mu$. Therefore, we must also have $|\lambda^{-1}| > \mu^{(1-2\nu)/2}$. Thus the conditions on λ are necessary for λC_φ to be hypercyclic.

Now, we turn to the proof that the conditions are also sufficient. Thus suppose that $\nu < 1/2$ and $\mu^{(1-2\nu)/2} < |\lambda| < \mu^{(2\nu-1)/2}$. Let X be the set of all holomorphic functions on a neighborhood of $\overline{\mathbb{D}}$ that vanish m times at 1, where m is to be determined later on. By Lemma 2.13, the set X is dense in $\mathcal{S}_\nu$. Let $f \in X$ be fixed. We have

$$\|\lambda^n C_{\varphi_n} f\|^2 = |\lambda|^{2n}|f(\varphi_n(0))|^2 + |\lambda|^{2n}\int_{\mathbb{D}}|f'(\varphi_n(z))|^2|\varphi_n'(z)|^2(1-|z|^2)^c dA(z), \quad (7)$$

where, as before, we have set $c = 1 - 2\nu$ and used Lemma 1.2.

Since $f(z) = (z-1)^m g(z)$, where $g(z)$ is holomorphic in a neighborhood of $\overline{\mathbb{D}}$, we have

$$|\lambda|^{2n}|f(\varphi_n(0))|^2 \leq \frac{|\lambda|^{2n}2^{2m}\mu^{2nm}}{(1+\mu^n)^{2m}} \max_{\overline{\mathbb{D}}} |g(z)|^2$$

that goes to zero as $n \to \infty$ if we choose $m > -\log|\lambda|/\log\mu$.

Hence, we only have to show that the second term in the right hand side of (7) goes to zero as $n \to \infty$. For this, we perform the change of variables $w = \varphi_n(z)$ in the integral in (7) and compute using (6) to obtain

$$\int_{\mathbb{D}}|f'(\varphi_n(z))|^2|\varphi_n'(z)|^2(1-|z|^2)^c dA(z) = \int_{\mathbb{D}}|f'(z)|^2(1-|\varphi_{-n}(z)|^2)^c dA(z)$$

$$= \mu^{nc}\int_{\mathbb{D}}\frac{4^c|f'(z)|^2(1-|z|^2)^c dA(z)}{|(\mu^n-1)z+\mu^n+1|^{2c}}.$$

As $f'(z) = (z-1)^{m-1}h(z)$, where $h(z)$ is holomorphic function on a neighborhood of $\overline{\mathbb{D}}$, we find that the above quantity is less than or equal to

$$\mu^{nc}\max_{\overline{\mathbb{D}}}|h(z)|^2\int_{\mathbb{D}}\frac{4^c|1-z|^{2m-2}(1-|z|^2)^c dA(z)}{|(\mu^n-1)z+\mu^n+1|^{2c}}. \quad (8)$$

Since $0 < \mu < 1$ and $c > 0$, an easy calculation shows that, for every positive integer n, the following inequality holds

$$\frac{|1-z|^{2c}}{|(\mu^n-1)z+\mu^n+1|^{2c}} \leq \frac{1}{(1-\mu)^{2c}} \quad (z \in \mathbb{D}).$$

We choose $m-1 > c$. Thus $|1-z|^{2m-2-2c} \leq 2^{2m-2-2c}$ for $z \in \mathbb{D}$. Upon putting everything together we find that (8) is less than or equal to

$$\frac{2^{2m-2}\mu^{nc}}{(1-\mu)^{2c}}\max_{\overline{\mathbb{D}}}|h(z)|^2\int_{\mathbb{D}}(1-|z|^2)^c.$$

Therefore, the second term in (7) is less than or equal to

$$C|\lambda|^{2n}\mu^{n(1-2\nu)},$$

where C is a constant independent of n. Now, the quantity above goes to zero as $n \to \infty$ because the condition on λ.

For the right inverse we take $S = \lambda^{-1}C_\varphi^{-1} = \lambda^{-1}C_{\varphi_{-1}}$ and the set Y will be the set of functions that are holomorphic on a neighborhood of $\overline{\mathbb{D}}$ and that vanish m times at -1. It is clear that Y is taken into itself by $\lambda^{-1}C_{\varphi_{-1}}$. As before, we can show that $\lambda^{-n}C_{\varphi_{-n}}$ tends pointwise to zero on Y whenever the hypothesis on λ is satisfied. Consequently, the hypercyclicity of λC_φ follows from the Hypercyclicity Criterion and, therefore, the conditions on λ and ν are also necessary and the proof is finished. $\square$

REMARK 3.6. For $\nu \leq 0$ the spectrum of a composition operator induced by a hyperbolic automorphism is the annulus $\varphi'(\eta)^{(1-2\nu)/2} \leq |\lambda| \leq \varphi'(\eta)^{(2\nu-1)/2}$ (see [Cw2] and [Cw3]). Thus λC_φ is hypercyclic if and only if $1/\lambda$ belongs to the interior of the spectrum. Again it seems that it is not known the spectrum of C_φ acting on $\mathcal{S}_\nu$ for $0 < \nu < 1/2$. Theorem 3.5 also suggests that the spectrum of C_φ for $0 < \nu < 1/2$ should maintain the same formula as for $\nu \leq 0$.

THE FAILURE OF THE SPECTRUM

As we have been seeing in the course of this work, the cut-off of cyclicity or hypercyclicity has always been determined by the spectrum, at least, when it is known. Even the spectrum has nearly determined the values of λ for which λC_φ is hypercyclic. This is no longer true in the present situation. However, it is still true that the spectrum can be used to determine the non cyclicity of C_φ on some of the spaces $\mathcal{S}_\nu$.

In the case that φ is a hyperbolic automorphism, the spectrum nearly determines the cut-off of cyclicity. Indeed, as noted by Zorboska [Zo3], for $\nu > 1/2$ the reproducing kernel

$$K_\alpha(z) = \sum_{n=0}^{\infty} \frac{\alpha^n}{(n+1)^{2\nu}} z^n \quad \text{with} \quad |\alpha| = 1,$$

belongs to $\mathcal{S}_\nu$. If α is a fixed point, then $C_\varphi^\star K_\alpha = K_{\varphi(\alpha)} = K_\alpha$. Now, a hyperbolic automorphism fixes two points on the boundary of the unit disk. Consequently, $C_\varphi^\star$ has an eigenvalue of multiplicity two. But it is known that the adjoint of a cyclic operator has no multiple eigenvalues (this fact follows from the obvious fact that $T - \lambda$ is also cyclic whenever T is, and the fact that the complement of the range of a cyclic operator is at most one dimensional). Therefore, C_φ cannot be cyclic. We stress here that, due to the lack eigenvalues for $C_\varphi^\star = C_{\varphi_{-1}}$, this method does not apply to the Dirichlet space ($\nu = 1/2$).

The situation is even worse when φ is a parabolic automorphism. We have the following Proposition which is due to Zorboska [Zo3] and that works for parabolic maps that are automorphisms or not. We include it for the sake of completeness.

THEOREM 3.7. *Let φ be a parabolic linear transformation that takes $\mathbb{D}$ into itself. Then C_φ is not cyclic on the weighted Dirichlet spaces $\mathcal{S}_\nu$ for $\nu > 3/2$.*

PROOF. By Proposition 1.4, there is only one fixed point α which is on $\partial \mathbb{D}$. Also we know that $\varphi'(\alpha) = 1$. Now, for $\nu > 3/2$ the reproducing kernel at 1

$$K_\alpha(z) = \sum_{n=0}^{\infty} \frac{1}{(n+1)^{2\nu}} z^n \quad \text{as well as} \quad K_\alpha^{(1)}(z) = \sum_{n=1}^{\infty} \frac{n}{(n+1)^{2\nu}} z^{n-1}$$

are in $\mathcal{S}_\nu$ for $\nu > 3/2$. On the other hand, $C_\varphi^\star K_\alpha = K_{\varphi(\alpha)} = K_\alpha$ and for any $f \in \mathcal{S}_\nu$

$$\langle C_\varphi^\star K_\alpha^{(1)}, f \rangle = \langle K_\alpha^{(1)}, C_\varphi f \rangle = f'(\varphi(\alpha))\varphi'(\alpha) = f'(\alpha) = \langle K_\alpha^{(1)}, f \rangle.$$

Hence, $C_\varphi^\star K_\alpha^{(1)} = K_\alpha^{(1)}$ and, thus, $C_\varphi^\star$ has an eigenvalue of multiplicity at least two. From this, it follows that C_φ is not cyclic for $\nu > 3/2$. The proof is complete. □

As we shall see in the next section, Theorem 3.7 is sharp for parabolic non automorphisms. But this is no longer true for parabolic automorphisms. The following proposition, that improves Theorem 3.7 when φ is a parabolic automorphism, states the best we can do using spectral methods.

PROPOSITION 3.8. *Let φ be a parabolic automorphism of the unit disk. If $\nu > 3/4$, then C_φ is not cyclic on $\mathcal{S}_\nu$; in fact, the closed linear span of any orbit has infinite codimension in $\mathcal{S}_\nu$.*

Recall that $\mathcal{S}_\nu^0$ denotes the space $\mathcal{S}_\nu$ modulo constants. To prove Proposition 3.8 we need a formula for the adjoint of C_φ acting on $\mathcal{S}_\nu^0$, where $0 < \nu < 1$. This formula will be valid for any automorphism of the unit disk and not just for parabolic automorphisms. Actually, it is an extension of Proposition 3.2.

THEOREM 3.9. *Let φ be an automorphism of the unit disk. If $0 < \nu < 1$, then the adjoint of C_φ acting on $\mathcal{S}_\nu^0$ is similar under a diagonal operator to C_φ^{-1} acting on $\mathcal{S}_{1-\nu}^0$.*

PROOF. We will use the equivalent norm given by Lemma 1.2 ($c = 1 - 2\nu$). First, for each $n \geq 1$ we set

$$w_n^2 = 2\int_0^1 r^{2n-1}(1-r^2)^c\,dr.$$

Then the norm of the monomials z^n ($n \geq 1$) is given by

$$\|z^n\| = \left(\int_\mathbb{D} n^2 |z|^{2n-2}(1-|z|^2)^c\,dA(z)\right)^{\frac{1}{2}} = \left(2\int_0^1 n^2 r^{2n-1}(1-r^2)^c\,dr\right)^{\frac{1}{2}} = nw_n.$$

We also write $u_n = z^n/\|z^n\| = z^n/(nw_n)$.

Let $f(z) = \sum_{k=1}^\infty a_k z^k$ be any function in $\mathcal{S}_\nu^0$. Also let $\sum_{k=0}^\infty b_k z^k$ be the Taylor expansion around the origin of φ^n. In the computation below the inner product is always that of $\mathcal{S}_\nu$. We have

$$\begin{aligned}\langle C_\varphi^\star f, u_n\rangle &= \frac{1}{nw_n}\langle f, C_\varphi z^n\rangle\\ &= \frac{1}{nw_n}\int_\mathbb{D} f'(z)n\overline{\varphi^{n-1}(z)\varphi'(z)}(1-|z|^2)^c\,dA(z)\\ &= \frac{2}{nw_n}\int_0^1 \sum_{k=1}^\infty k^2 a_k \bar b_k r^{2k-1}(1-r^2)^c\,dr\\ &= \frac{1}{nw_n}\sum_{k=1}^\infty k^2 w_k^2 a_k \bar b_k\\ &= \frac{2}{nw_n}\int_0^1 \sum_{k=1}^\infty k^3 w_k^2 a_k \bar b_k r^{2k-1}\,dr. \end{aligned} \qquad (9)$$

We consider the diagonal operator defined by

$$Wf(z) = \sum_{k=1}^\infty k w_k^2 a_k z^k.$$

This operator defines an isometry from $\mathcal{S}_\nu^0$ onto $\mathcal{S}_{1-\nu}^0$. We set $g = Wf$ and we can write the integral in (9) as an integral over the whole disk $\mathbb{D}$

$$\frac{1}{nw_n} \int_{\mathbb{D}} g'(z) n \overline{\varphi(z)^{n-1} \varphi'(z)} \, dA(z).$$

As in the proof of Proposition 3.2, the change of variables $w = \varphi(z)$ in the integral above yields

$$\frac{1}{nw_n} \int_{\mathbb{D}} g'(\varphi_{-1}(z)) \varphi'_{-1}(z) n \bar{z}^{n-1} \, dA(z).$$

The last integral is the n-th coefficient of $h = g \circ \varphi_{-1}$ multiplied by n. If we denote this coefficient by c_n and the n-th coefficient of $W^{-1}h$ by d_n, we can continue the calculation as follows

$$\frac{1}{w_n} c_n = nw_n d_n = \frac{1}{nw_n} \langle W^{-1}h, z^n \rangle = \langle W^{-1} C_{\varphi_{-1}} W f, u_n \rangle.$$

Thus we have shown that

$$C_\varphi^\star = W^{-1} C_{\varphi_{-1}} W \tag{10}$$

and the result follows. $\square$

The eigenfunctions of C_φ with φ a parabolic automorphism.

We also need to know the eigenvalues of C_φ. Recall from the proof of Theorem 3.3 that we may assume that φ has the following expression

$$\varphi(z) = \frac{(2-a)z + a}{-az + 2 + a} \qquad \text{with } a \neq 0 \text{ and } \Re a = 0. \tag{11}$$

Carl Cowen showed that the spectrum of C_φ acting on the Hardy space is the circle $\{e^{-at} : t \geq 0\} \cup \{0\}$ (see [CM, Thm. 6.1]). In particular, using the expression in (11) for φ, it is easy to check that for each non negative number $t \geq 0$ the following inner function

$$e_t(z) = \exp\left[t \frac{z+1}{z-1}\right]$$

is an eigenfunction of C_φ corresponding to the eigenvalue e^{-at}. Since these functions are inner, they are in the Hardy space $\mathcal{H}^2$. On the other hand, $e_0(z) = 1$ is in any of the $\mathcal{S}_\nu$ spaces. However, this is not true for $t > 0$. Actually, from the work in [NS] it can be easily deduced that they are in $\mathcal{S}_\nu$ if and only if $\nu < 1/4$. Here, we provide an alternative and simpler proof of this fact.

PROPOSITION 3.10. *Suppose that $t > 0$. Then the function $e_t(z)$ is in $\mathcal{S}_\nu$ if and only if $\nu < 1/4$.*

PROOF. First observe that it is enough to show the result for $0 < \nu < 1/4$. We use the norm furnished by Lemma 1.2 ($c = 1 - 2\nu$ and $l = 0$). The function $e_t(z)$ is in $\mathcal{S}_\nu$ if and only if

$$\int_{\mathbb{D}} \left|\exp\left[t\frac{z+1}{z-1}\right]\right|^2 t^2 |\sigma'(z)|^2 (1-|z|^2)^c dA(z) < \infty.$$

The change of variables

$$w = \sigma(z) = \frac{i(1+z)}{1-z}$$

that takes $\mathbb{D}$ onto the upper half plane yields

$$t^2 \int_\Pi |e^{itw}|^2 (1-|\sigma^{-1}(w)|^2)^c \, dA(w) = \int_0^\infty \int_0^\infty 2t^2 e^{-2ty} \left[\frac{4y}{x^2+(y+1)^2}\right]^c dx dy.$$

By comparing with an appropriate series, one easily checks that the integral on the right hand side above is convergent if and only if $2c > 1$, that is, $\nu < 1/4$. $\square$

PROOF OF PROPOSITION 3.8. By the Comparison Principle we may suppose that $3/4 < \nu < 1$ and by Lemma 2.15, it is enough to prove the result for $\mathcal{S}_\nu^0$.

As in [CM, Thm. 7.5], we consider for each $t \geq 0$ the function $e_t(z)$. We have already noticed that

$$e_t(\varphi(z)) = e^{-at} e_t(z).$$

Thus $e_t(z)$ with $t > 0$ is an eigenfunction for C_φ corresponding to the eigenvalue e^{-at} and whenever $\nu < 1/4$. Since each point on the unit circle has infinitely many representations of this form, every point of the unit circle is an eigenvalue of infinite multiplicity for C_φ.

Now we consider C_φ acting on $\mathcal{S}_\nu^0$. The functions

$$d_t = e_t(z) - e^{-t} \quad (t > 0)$$

satisfy

$$C_\varphi d_t = e_t(\varphi(z)) - e^{-t} - (e_t(\varphi(0)) - e^{-t}) = e^{-at} e_t(z) - e^{-at} e^{-t} = e^{-at} d_t.$$

Thus the functions d_t with $t > 0$ are eigenfunctions of infinite multiplicity of C_φ acting on $\mathcal{S}_\nu^0$.

The representation for the adjoint in (10) shows that $C_{\varphi_{-1}}^\star = W^{-1} C_\varphi W$. Now, the fact that for $t > 0$ the function $d_t(z)$ is in $\mathcal{S}_{1-\nu}^0$ for $3/4 < \nu < 1$ implies that the function

$$f_t(z) = W^{-1} d_t(z),$$

is in $\mathcal{S}_\nu$ for $3/4 < \nu < 1$. Hence, f_t is an eigenfunction of infinite multiplicity for $C_{\varphi_{-1}}^\star$. Therefore, as in the proof of Theorem 2.6, it follows that $C_{\varphi_{-1}}$ is strongly non cyclic. Finally, as φ and φ_{-1} are interchangeable, the result follows. $\square$

CYCLICITY

In this section we will show by different methods that composition operators induced by non elliptic automorphisms are not cyclic when they act on the Dirichlet space. This result along with Theorem 3.3 and 3.5 shows that the cut-off of cyclicity of non-elliptic automorphisms is at the Dirichlet space.

Recall that $\mathcal{D}_0$ denotes the space of all Dirichlet functions modulo constant functions. By Proposition 3.2, composition operators induced by parabolic or hyperbolic automorphisms are unitary operators on $\mathcal{D}_0$. This makes these operators quite manageable and allows us to prove that they are not cyclic.

For the work in this section it will be convenient to transfer the problem to the upper half plane.

The Dirichlet space of the upper half plane.

Let Π denote the upper half plane of the complex plane. The Dirichlet space of the upper half plane $\mathcal{D}_\pi$ consists of those functions holomorphic on the upper half plane Π for which

$$\|F\|_{\mathcal{D}_\pi}^2 = \frac{1}{\pi} \int_\Pi |F'(z)|^2 \, dA(z) = \frac{1}{\pi} \int_{-\infty}^{\infty} \int_0^{\infty} |F'(x+iy)|^2 \, dx dy$$

is finite. If we identify functions that differ by a constant, $\mathcal{D}_\pi$ becomes a Hilbert space. The change of variables

$$w = \sigma(z) = i\frac{1+z}{1-z},$$

that takes $\mathbb{D}$ onto Π, shows that $F \in \mathcal{D}_\pi$ if and only if $F \circ \sigma \in \mathcal{D}_0$. Hence, the following composition operator

$$C_\sigma : \mathcal{D}_\pi \to \mathcal{D}_0$$

induces an isometric isomorphism between $\mathcal{D}_\pi$ and $\mathcal{D}_0$.

The Hardy space of the upper half plane.

We need a version of the Paley-Wiener Theorem for $\mathcal{D}_\pi$. But, first we recall the classical one in the Hardy space of the upper half plane.

The Hardy space of the upper half plane $\mathcal{H}^2(\Pi)$ is defined as the space of holomorphic functions on the upper half plane for which the norm

$$\|f\|_{\mathcal{H}^2(\Pi)}^2 = \sup_{0<y<\infty} \frac{1}{\sqrt{2\pi}} \int_{-\infty}^{\infty} |f(x+iy)|^2 \, dx$$

is finite. A theorem of Paley and Wiener (see [Ru1, p. 372]) asserts that $\mathcal{H}^2(\Pi)$ is isometrically isomorphic, under the Fourier transform, to the space of functions which are square integrable on the real positive axis with respect to the Lebesgue measure normalized by $\sqrt{2\pi}$, that is, $L^2(\mathbb{R}^+, dt/\sqrt{2\pi})$. To each function $F \in \mathcal{H}^2(\Pi)$ there corresponds $f \in L^2(\mathbb{R}^+, dt/\sqrt{2\pi})$ such that

$$F(z) = \frac{1}{\sqrt{2\pi}} \int_0^{\infty} f(t) e^{itz} \, dt \qquad (z \in \Pi)$$

and

$$\|F\|_{\mathcal{H}^2(\Pi)}^2 = \|f\|_{L^2(\mathbb{R}^+)}^2 = \frac{1}{\sqrt{2\pi}} \int_0^{\infty} |f(t)|^2 \, dt.$$

Moreover, Placherel's Theorem asserts that the Fourier transform of F, which we denote by $\widehat{F}$, coincides with f. In particular, $\widehat{F}(x) = 0$ for almost every $x < 0$.

A Paley-Wiener Theorem for $\mathcal{D}_\pi$.

In the same vein as the Paley-Wiener for $\mathcal{H}^2(\Pi)$, there is a result due to Higdon [Hi] for the Dirichlet space $\mathcal{D}_\pi$.

THEOREM 3.11. *For each function F in $\mathcal{D}_\pi$ there corresponds a function $f \in L^2(\mathbb{R}^+, tdt/(2\pi))$ such that*

$$\|F\|_{\mathcal{D}_\pi}^2 = \frac{1}{2\pi} \int_0^{\infty} t|f(t)|^2 \, dt,$$

and this correspondence is an isometry from $\mathcal{D}_\pi$ onto $L^2(\mathbb{R}^+, tdt/(2\pi))$.

In fact, for $F \in \mathcal{H}^2(\Pi) \cap \mathcal{D}_\pi$ the relation between F and f is given by the Fourier transform

$$f(t) = \frac{1}{\sqrt{2\pi}} \int_{-\infty}^{\infty} F(z) e^{-itz} \, dz, \qquad (12)$$

where the last integral is taken over any horizontal line in Π, that is, if $\widehat{F}$ denotes the Fourier transform of F, then $\hat{F} = f$.

The proof of Theorem 3.11 is based on the Paley-Wiener Theorem for the Hardy space $\mathcal{H}(\Pi)$, Plancherel's Theorem, and a density argument as well as in the properties of the Fourier transform. The density argument is based on the following lemma

LEMMA 3.12. *The subspace $\mathcal{H}^2(\Pi) \cap \mathcal{D}_\pi$ is dense in $\mathcal{D}_\pi$.*

For a proof of the lemma above see [Hi]. It will also be needed in the proofs of Theorems 3.13 and 3.14.

Now, we are in position to prove the non cyclicity of C_φ whenever φ is a non elliptic automorphism.

The parabolic automorphism.

THEOREM 3.13. *Let φ be a parabolic automorphism of the unit disk. Then C_φ is not cyclic on the Dirichlet space; in fact, the closed linear span of any orbit has infinite codimension in $\mathcal{D}$. In particular, the same is true for $\mathcal{S}_\nu$ whenever $\nu \geq 1/2$.*

PROOF. By the Comparison Principle it is sufficient to prove the result for $\mathcal{D}$ and by Lemma 2.15 it is enough to prove that C_φ is not cyclic on $\mathcal{D}_0$. As usual, we may assume that the fixed point of φ is 1. Now, C_φ is similar under

$$C_\sigma : \mathcal{D}_\pi \to \mathcal{D}_0$$

to

$$C_\tau : \mathcal{D}_\pi \to \mathcal{D}_\pi,$$

where $\tau(z) = z + a$ and a is a non-zero real number; that is, $C_\varphi = C_\sigma C_\tau C_{\sigma^{-1}}$.

To show that C_τ is not cyclic on $\mathcal{D}_\pi$ we need a further similarity. First, suppose that $F \in \mathcal{H}^2(\Pi) \cap \mathcal{D}_\pi$. Then the properties of the Fourier transform show that

$$(F(z+a))\hat{\,} = e^{iat} \hat{F}(t).$$

Let $\phi : \mathbb{R}^+ \to \mathbb{R}^+$ be defined by $\phi(t) = e^{iat}$. By Lemma 3.12, $\mathcal{H}^2(\Pi) \cap \mathcal{D}_\pi$ is dense in $\mathcal{D}_\pi$. Consequently, the operator C_τ on $\mathcal{D}_\pi$ is similar under the Fourier Transform to the multiplication operator

$$M_\phi : L^2(\mathbb{R}^+, t\,dt/(2\pi)) \longrightarrow L^2(\mathbb{R}^+, t\,dt/(2\pi))$$

defined by the pointwise multiplication $M_\phi g(t) = \phi(t) g(t)$. Thus it is enough to prove the statement of the theorem for M_ϕ. Therefore, consider any function $f \in L^2(\mathbb{R}^+, t\,dt/(2\pi))$. We have

$$\text{span}\,\{M_\phi^n f : n = 0, 1, 2, \dots\} = \{pf : \text{where } p \text{ is a polynomial in } e^{iat}\}.$$

First, suppose that f vanishes on a set $A \subset \mathbb{R}^+$ of positive measure. Since the function pf also vanishes on A for any polynomial p, we cannot approximate any function that is different from zero on A. Therefore, f cannot be cyclic and, moreover, the closed linear span of the orbit of f has infinite codimension. In this case, the closed linear span of the orbit of f lies on a non-trivial reducing subspace of M_ϕ, that is, an invariant subspace whose orthogonal is also invariant.

Now, suppose that f is different from zero in any set of positive measure. In this case, the non-cyclicity of f is due to the fact that any polynomial in e^{iat} is of period $\omega = 2\pi/a$. Let $\mathcal{X}_{[c,d]}$ denote the characteristic function of the interval $[c,d]$. We set
$$h = \mathcal{X}_{[0,\omega]}$$
that belongs to $L^2(\mathbb{R}^+, t\,dt/(2\pi))$. Suppose that there is a sequence $\{p_n\}$ of polynomials in e^{iat} such that $\{p_n f\}$ tends to h in the norm of $L^2(\mathbb{R}^+, t\,dt/(2\pi))$. Then there is a subsequence $\{p_{n_k} f\}$ that tends pointwise to h almost everywhere. Therefore, since f is different from zero almost everywhere, $p_{n_k}(t)$ tends to 0 for almost every $t \in \mathbb{R}^+ \setminus [0,\omega]$ and to $1/f(t)$ for almost every $t \in [0,\omega]$. It follows, by periodicity of the elements of the sequence $\{p_{n_k}\}$, that $1/f(t) = 0$ for almost every $t \in [0,\omega]$; a contradiction. Thus f cannot be cyclic. Finally, the infinite set of functions
$$\mathcal{X}_{[k\omega,(k+1)\omega]} \qquad (k \geq 1)$$
is linearly independent and a similar argument as above shows that none of them can be approximated by a sequence $\{p_n f\}$, where $\{p_n\}$ is a sequence of polynomials in e^{iat}. The proof is complete. $\square$

Theorems 3.13 and 3.3 determine the cut-off of hypercyclicity and cyclicity of composition operators induced by parabolic automorphisms of $\mathbb{D}$.

The hyperbolic automorphism.

THEOREM 3.14. *Let φ be a hyperbolic automorphism of the unit disk. Then C_φ is not cyclic on the Dirichlet space; in fact, the closed linear span of any orbit has infinite codimension in $\mathcal{D}$. In particular, the same is true for $\mathcal{S}_\nu$ whenever $\nu \geq 1/2$.*

PROOF. Again, by the Comparison Principle and Lemma 2.15 it is enough to prove the result for $\mathcal{D}_0$. Without loss of generality we assume that the fixed points of φ are -1 and 1. By means of
$$C_\sigma : \mathcal{D}_\pi \to \mathcal{D}_0$$
we transfer the problem to $\mathcal{D}_\pi$ and we find that
$$C_\varphi = C_{\sigma^{-1}} C_{\lambda z} C_\sigma \quad \text{where } 0 < \lambda < 1.$$

Therefore, it is enough to show that $C_{\lambda z}$ is not cyclic on $\mathcal{D}_\pi$. Using the properties of the Fourier transform, one easily checks that for $F \in \mathcal{H}^2(\Pi) \cap \mathcal{D}_\pi$ the formula
$$(F(\lambda z))\hat{\,} = (1/\lambda)\hat{F}(t/\lambda)$$
holds. Again, the fact that $\mathcal{H}^2(\Pi) \cap \mathcal{D}_\pi$ is dense in $\mathcal{D}_\pi$ shows that $C_{\lambda z} : \mathcal{D}_\pi \to \mathcal{D}_\pi$ is similar to
$$T_\lambda : L^2(\mathbb{R}^+, t\,dt/(2\pi)) \longrightarrow L^2(\mathbb{R}^+, t\,dt/(2\pi))$$
defined by $T_\lambda f(t) = (1/\lambda)f(t/\lambda)$. This time we need a further Fourier transform that yields to one more similarity. But before to do this we need another similarity. The identity
$$\frac{1}{2\pi}\int_0^\infty |f(t)|^2 \, t\,dt = \frac{1}{2\pi}\int_0^\infty |tf(t)|^2 \, \frac{dt}{t}$$

shows that the map $f(t) \to tf(t)$ induces an isometric isomorphism from the space $L^2(\mathbb{R}^+, tdt/(2\pi))$ onto $L^2(\mathbb{R}^+, dt/(2\pi t))$. Under this isometry T_λ is similar to the composition operator

$$C_{t/\lambda} : L^2(\mathbb{R}^+, dt/(2\pi t)) \longrightarrow L^2(\mathbb{R}^+, dt/(2\pi t))$$

defined by $C_{t/\lambda} f(t) = f(t/\lambda)$. Now, observe that $dt/(2\pi t)$ is the Haar measure corresponding to the multiplicative locally compact Abelian group $\mathbb{R}^+$, that is, the measure μ that is uniquely defined except for a positive scalar multiple and that satisfies $\mu(rA) = \mu(A)$ for any positive r and any measurable set A. The group of characters corresponding to this measure is formed by

$$\gamma_t(x) = x^{it} = e^{it \log x} \quad (t \in \mathbb{R})$$

and thus the dual group is the additive group of real numbers (see [Ru2, Section 1.2]). This time the Fourier transform for $f \in L^2(\mathbb{R}^+, dt/(2\pi t))$ is defined as

$$\hat{f}(t) = \frac{1}{\sqrt{2\pi}} \int_0^\infty f(x) x^{-it} \frac{dx}{x}.$$

Observe that $\hat{f}$ is defined on $\mathbb{R}$. By Plancherel's Theorem (see [Ru2, Thm. 1.6.1], for instance) the Fourier transform defines an isometry from $L^2(\mathbb{R}^+, dt/(2\pi t))$ onto $L^2(\mathbb{R}, m)$, where m denotes the Lebesgue measure normalized by $\sqrt{2\pi}$.

Now, for $g(x) \in L^2(\mathbb{R}^+, dt/(2\pi t))$, we have

$$\int_0^\infty g(x/\lambda) x^{-it} \frac{dx}{x} = \lambda^{-it} \int_0^\infty g(x) \frac{dx}{x}.$$

Therefore, if write $\phi(t) = \lambda^{-it}$, we find that the composition operator $C_{t/\lambda}$ acting on $L^2(\mathbb{R}^+, dt/(2\pi t))$ is similar under the last Fourier transform to the multiplication operator M_ϕ acting on $L^2(\mathbb{R}, m)$. Observe that any polynomial in ϕ is of period $\omega = -2\pi/\log \lambda$. Therefore, similar arguments to those used in the parabolic automorphism case show that the orbit generated by any function has infinite codimension, which is the desired result. □

Theorems 3.14 and 3.5 determine the cut-off of hypercyclicity and cyclicity of composition operators induced by hyperbolic automorphisms of $\mathbb{D}$. Therefore, we have completed the proof of Theorem 3.1.

REMARK 3.15. We end the chapter by noting that the techniques we used in Chapter 2 cannot be used to prove the cyclicity of C_φ for φ a non elliptic automorphism and $\nu < 1/2$. The point is that in the present situation the sequence $\{\varphi_n(z)\}$, for any $z \in \mathbb{D}$, forms a Blaschke sequence.

CHAPTER 4

THE PARABOLIC NON AUTOMORPHISM

In this chapter we deal with composition operators induced by a parabolic non automorphism, which have the most exotic cyclic behavior. The following theorem will be proved

THEOREM 4.1. *Let φ be a parabolic non automorphism that takes the unit disk into itself. Then C_φ acting on the weighted Dirichlet space S_ν is cyclic if and only if $\nu \leq 3/2$. In addition, λC_φ is never hypercyclic for any ν and for any λ.*

The cyclic part of Theorem 4.1 shows exactly the cut-off of cyclicity of composition operators induced by parabolic non automorphisms and answers a question posed by Zorboska. With respect to the hypercyclic part of Theorem 4.1, Shapiro [Sh3] has previously proved that λC_φ acting on $\mathcal{H}^2$ is never hypercyclic. The extension of this result to weighted Dirichlet spaces S_ν is done by mean of a recursion argument that involves a special complete orthogonal system of $L^2(\mathbb{R}^+, dt/t)$ and functional Hilbert spaces. In Chapter 5, Theorem 5.2, we will strengthen Theorem 4.1 by proving that C_φ is not even supercyclic.

CYCLICITY

First we dispense with the cyclicity. We will see that the cyclicity of C_φ is closely related to the growth conditions of the sequence of zeros of functions in S_ν.

THEOREM 4.2. *Suppose that φ is a parabolic non-automorphism that takes $\mathbb{D}$ into itself. Then C_φ is cyclic on the weighted Dirichlet space S_ν if and only if $\nu \leq 3/2$.*

PROOF. By Theorem 3.6 we already know that C_φ is not cyclic for $\nu > 3/2$. Thus we need only prove that C_φ is cyclic for $\nu \leq 3/2$. According to Proposition 1.4, a parabolic non-automorphism that takes $\mathbb{D}$ into itself must have its fixed point on $\partial \mathbb{D}$. Now, φ is conjugate under an appropriate disk automorphism to a self map having 1 as its fixed point. Therefore, since cyclicity is invariant under similarity we may assume from the beginning that the fixed point is 1. By conjugating with

$$\sigma(z) = \frac{i(1+z)}{1-z}$$

that takes $\mathbb{D}$ onto the upper half plane and 1 to ∞, we obtain that φ is conjugate to a translation on the upper half plane. Therefore,

$$w = \varphi(z) \equiv \frac{1+w}{1-w} = \frac{1+z}{1-z} + a, \qquad (1)$$

where $\Re a > 0$ because φ is not an automorphism of the unit disk. From equation (1) it is easy to compute the iterates of φ

$$\varphi_n(z) = \frac{(2-na)z + na}{-naz + 2 + na}. \tag{2}$$

Let us prove that for $\nu \leq 3/2$ the operator C_φ is cyclic on $\mathcal{S}_\nu$. By the Comparison Principle, it is enough to consider the case in which $\nu = 3/2$. We will prove that the identity map, $u(z) = z$ defined on $\mathbb{D}$, is a cyclic vector for C_φ.

Using (2) and a straightforward decomposition we have

$$\varphi_n = \bar{\gamma}_n + \bar{\alpha}_n K_{w_n}, \tag{3}$$

where

$$\bar{\gamma}_n = \frac{na-2}{na}, \quad \bar{\alpha}_n = \frac{4}{na(na+2)}, \quad \bar{w}_n = \frac{na}{2+na}$$

and $K_{w_n} = (1 - \bar{w}_n z)^{-1}$. All quantities above are well defined because $\Re a > 0$. Observe that $w_n \in \mathbb{D}$. Thus K_{w_n} is the reproducing kernel at w_n in the Hardy space $\mathcal{H}^2$. For $f(z) = \sum_{n=0}^\infty a_n z^n \in \mathcal{S}_{3/2}$ we consider the function

$$g(z) = \sum_{n=0}^\infty (n+1)^3 a_n z^n$$

which clearly belongs to $\mathcal{S}_{-3/2}$. Now, for $K_w = (1-\bar{w}z)^{-1} = \sum_{n=0}^\infty \bar{w}^n z^n$, the reproducing kernel (in $\mathcal{H}^2$) at the point $w \in \mathbb{D}$, we have

$$\langle f, K_w \rangle = \sum_{n=0}^\infty a_n w^n (n+1)^3 = g(w).$$

Now suppose that $f \in \mathcal{S}_\nu$ is orthogonal to the orbit of u under C_φ. To show the cyclicity of C_φ it is enough to prove that f is the zero function. We have

$$\langle f, \varphi_n \rangle = 0 \quad \text{for } n = 0, 1, \ldots,$$

where $\varphi_0 = u$. We claim that f is also orthogonal to the constant functions. To show this, it is enough to prove that $\{\varphi_n\}$ tends weakly to 1, since in such a case we have

$$0 = \lim_n \langle f, \varphi_n \rangle = \langle f, 1 \rangle = f(0).$$

Now, obviously $\{\varphi_n\}$ tends uniformly on compact subsets of $\mathbb{D}$ to the constant function 1. Thus to prove that $\{\varphi_n\}$ tends weakly to 1, it is enough to show that $\{\varphi_n\}$ is bounded in the norm of $\mathcal{S}_{3/2}$. Toward this end, we use the equivalent norm furnished by Lemma 1.2 (in this case l=1). Since $\varphi_n(0)$ tends to 1 and $\varphi'_n(0) \to 0$ as n tends to ∞, it is sufficient to prove that

$$\int_\mathbb{D} |\varphi''_n(z)|^2 \, dA(z)$$

is uniformly bounded. We compute φ''_n by using (2). Thus the above integral is equal to

$$\int_\mathbb{D} \frac{64 n^2 |a|^2}{|2+na-naz|^6} \, dA(z) = \frac{64 n^2 |a|^2}{\pi |2+na|^6} \int_0^1 r \, dr \int_0^{2\pi} \frac{d\theta}{|1 - na/(2+na) r e^{i\theta}|^6}. \tag{4}$$

We set
$$c = c(n) = \frac{-na}{2+na}$$
that belongs to $\mathbb{D}$ for every non negative integer. Then the change of variables $z = e^{i\theta}$ in the second integral in (4) yields
$$\int_0^{2\pi} \frac{d\theta}{|1+cre^{i\theta}|^6} = \int_{|z|=1} \frac{z^2 dz}{i(crz^2 + (1+|c|^2r^2)z + \bar{c}r)^3}.$$

The integrand in the right-hand side above has triples poles at $-\bar{c}r$ and $-1/(cr)$, but only $-\bar{c}r$ belongs to $\mathbb{D}$. Thus applying the Residue Theorem we obtain that the last integral is equal to
$$2\pi \frac{1 + |c|^4 r^4 + 4|c|^2 r^2}{(1-|c|^2 r^2)^5}.$$

Upon substituting in (4) we obtain
$$\frac{128 n^2 |a|^2}{|2+na|^6} \int_0^1 \frac{r + |c|^4 r^5 + 4|c|^2 r^3}{(1-|c|^2 r^2)^5} \, dr. \tag{5}$$

Since $|c(n)| < 1$, we see that
$$\frac{r + |c|^4 r^5 + 4|c|^2 r^3}{(1+|c|r)^5}$$
remains bounded independently of n. Therefore, there is a constant M such that (5) is less or equal than
$$\frac{Mn^2}{|2+na|^6} \int_0^1 \frac{dr}{(1-|c|r)^5} = \frac{Mn^2}{4|c||2+na|^6} \left(\frac{1}{(1-|c|)^4} - 1 \right)$$
$$< \frac{Mn^2}{4|c||2+na|^2(|2+na|-|na|)^4}$$
$$= \frac{Mn^2(|2+na|+|na|)^4}{4|c||2+na|^2(4+4n\Re a)^4}.$$

Since $\Re a > 0$, the last quantity remains bounded as n tends to ∞ (this would not be true if $\Re a = 0$ which is the case of the parabolic automorphism). Thus $\{\varphi_n\}$ tends weakly to 1.

In conclusion, we have $f(0) = 0$. Using this fact along with the orthogonality of f and φ_n, we find that
$$0 = \langle f, \varphi_n \rangle = \langle f, \bar{\gamma}_n + \bar{\alpha}_n K_{w_n} \rangle = \gamma_n \langle f, 1 \rangle + \alpha_n \langle f, K_{w_n} \rangle = \alpha_n g(w_n),$$
so that $g(w_n) = 0$ for every n. Therefore, g vanishes on the sequence of points $\{w_n\}$. In addition, we have
$$1 - |w_n|^2 = \frac{4(1+n\Re a)}{4 + 4n\Re a + n^2|a|^2} \geq \text{const.} \frac{1}{n}.$$

Hence, $\sum_{n=1}^{\infty} 1 - |w_n| = \infty$, that is, $\{w_n\}$ is not a Blaschke sequence.

Now, observe that $\bar{w}_n = \varphi_n(0)$. Since $\Re a > 0$, the sequence $\{\varphi_n(0)\}$ tends non-tangentially to 1. This provides a constant $C > 0$ such that for each positive integer n
$$1 - |\varphi_n(0)| \geq C|1 - \varphi_n(0)|.$$

This implies that the sequence $\{w_n\}$ lies inside a Stolz angle and, therefore, it lies on the disk centered at $1/2$ and of radius $1/2$ for n large enough. We claim that all these conditions on the zero sequence of g which is in $\mathcal{S}_{-3/2}$ imply that g is identically 0 on $\mathbb{D}$ and, therefore, so is f. Thus once we have proved Lemma 4.3 below, the proof of the theorem will be completed. □

Zeros of weighted Dirichlet functions.

LEMMA 4.3. *Let $g \neq 0$ be in $\mathcal{S}_\nu$ where $\nu < 0$. If g vanishes on a sequence $\{w_n\}$ that lies on $D(1/2, 1/2)$, the disk centered at $1/2$ and of radius $1/2$, then $\{w_n\}$ must be a Blaschke sequence, that is,*

$$\sum_{n=1}^{\infty}(1-|w_n|) < \infty.$$

Of course, for $\nu \geq 0$ the space $\mathcal{S}_\nu$ is contained in $\mathcal{H}^2$ and the conclusion of the lemma is true for any sequence contained in $\mathbb{D}$. The difference between the statement of Lemma 4.3 and that of Theorem 4 in [SS] is that the sequence $\{w_n\}$ is allowed to lie on a disk tangent to $\mathbb{D}$ and not just on the interval $[0, 1]$. However, the proof only differs in details.

PROOF. Let us denote by $\log^+ |g(z)| = \max\{\log |g(z)|, 0\}$. For $0 < r \leq 1/2$, we also denote by C_r the circle centered at $1/2$ and of radius r. We will prove that

$$\int_{C_r} \log^+ |g(z)| \, |dz| \leq \text{const.} \qquad \text{for } 0 < r \leq 1/2. \tag{6}$$

That is, g is a function of bounded characteristic relative to the disk $D(1/2, 1/2)$. Thus a standard argument based on Jensen's formula (see [Sh2, pp. 120–121], for instance) shows that $\{w_n\}$ must be a Blaschke sequence.

First, we show that if $g(z) = \sum_{k=0}^{\infty} a_k z^k \in \mathcal{S}_\nu$ with $\nu < 1/2$, then there is a constant $C > 0$ such that the following growth estimate is satisfied

$$|g(z)| \leq \frac{C}{(1-|z|^2)^{1/2-\nu}} \qquad (z \in \mathbb{D}). \tag{7}$$

This estimate follows as the growth estimate for the first derivative. By applying the Cauchy-Schwarz inequality to the power series representation of g and Stirling's formula we find, for each $z \in \mathbb{D}$, that

$$|g(z)| \leq \sum_{n=0}^{\infty} |a_n|(n+1)^\nu \frac{|z|^n}{(n+1)^\nu}$$

$$\leq \left(\sum_{n=0}^{\infty} |a_n|^2 (n+1)^{2\nu}\right)^{1/2} \left(\sum_{n=0}^{\infty} \frac{|z|^{2n}}{(n+1)^{2\nu}}\right)^{1/2}$$

$$\approx \frac{\|g\|}{(1-|z|^2)^{1/2-\nu}},$$

which is the desired estimate.

From (7) we deduce that for any fixed $0 < \delta < 1/2$, there is another constant M such that

$$|g(z)| \leq M \exp \frac{1}{(1-|z|)^\delta}.$$

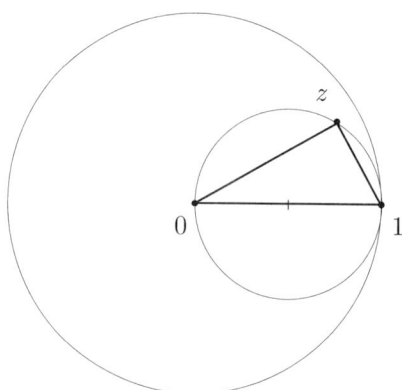

Figure 2

Let $z \in C_r$ and let $t = t(z)$ be the arc length along C_r from the point z to the point $1/2 + r$. A little trigonometry shows that $t^2/(1 - |z|)$ remains bounded on the circle $C_{1/2}$ and, therefore, so does for any circle C_r for $0 < r \leq 1/2$ (see Figure 2).

Hence, g as a function of t satisfies $\log|g(t)| \leq Lt^{-2\delta}$ for some constant L from which (6) follows. The proof of the lemma and, therefore, that of Theorem 4.2 is concluded. □

REMARK 4.4. We stress here that if $\nu < 3/2$, then $\{\varphi_n\}$ tends to 1 not only weakly, but also in the norm of $\mathcal{S}_\nu$. Thus a similar argument would work to show that φ induces a cyclic composition operator in $\mathcal{S}_\nu$ without using the Comparison Principle.

When C_φ fails to be cyclic on $\mathcal{S}_\nu$ one can ask if it is strongly non cyclic. The parabolic non-automorphism case is the only one for which the answer is negative. In fact, the cut-off at $3/2$ is quite mild. If we consider $\mathcal{S}_\nu^0$, the space $\mathcal{S}_\nu$ modulo constants, then we do not need to prove that the C_φ-orbit of $u(z) = z$ is orthogonal to the constant functions. Therefore, from the proof of Theorem 4.2, we have

COROLLARY 4.5. *If φ is a parabolic non-automorphism that takes $\mathbb{D}$ into itself, then C_φ is cyclic on any of the $\mathcal{S}_\nu^0$ spaces.*

NON HYPERCYCLICITY

Now we study the hypercyclicity of λC_φ. Throughout the rest of this section we will suppose that $\{\varphi_n\}$ satisfies equation (2) for some complex number a with $\Re a > 0$.

An appetizer.

First, we will show that in the Bergman space (and therefore, by the Comparison Principle in the Hardy space) C_φ is not hypercyclic. The result for the Hardy space is due to Bourdon and Shapiro. Our proof is simpler than that of Bourdon and Shapiro (see [BS2, pp. 28-30] or [Sh, pp. 115-117]). We take any function $f \in \mathcal{S}_{-1/2}$. The growth estimate for the first derivative, Proposition 2.16, provides a constant C such that the inequality below holds

$$\begin{aligned}|\langle C_\varphi^n f, z\rangle| &= |\langle f \circ \varphi_n, z\rangle| \\ &= \frac{1}{2}|f'(\varphi_n(0))||\varphi_n'(0)| \\ &\leq \frac{C}{(1-|\varphi_n(0)|^2)^2}\frac{2}{|2+na|^2} \\ &= \frac{C|2+na|^2}{8|1+n\Re a|^2}.\end{aligned}$$

Since $\Re a > 0$, the above quantity remains bounded. Therefore, we can apply Lemma 2.15 to see that C_φ is not hypercyclic on $\mathcal{S}_{-1/2}$.

The inner product of elements of the orbit against the function $g(z) = z$ does not work for spaces strictly greater than the Bergman space because the growth estimate for the derivative is not appropriate. The remainder of this section is devoted to proving the corresponding result for any of the $\mathcal{S}_\nu$ spaces.

THEOREM 4.6. *Suppose that φ is a parabolic non-automorphism that takes the unit disk into itself and λ is any complex number. Then λC_φ is not hypercyclic on any of the weighted Dirichlet spaces $\mathcal{S}_\nu$. In particular, C_φ is not hypercyclic on any of the $\mathcal{S}_\nu$ spaces.*

The spectrum of C_φ.

The proof of Theorem 4.6 will require several lemmas and depends, in an essential way, on the eigenfunctions of C_φ. Carl Cowen showed that the spectrum of C_φ acting on the Hardy space is $\{e^{-at} : t \geq 0\} \cup \{0\}$ (see [CM, Thm. 6.1]). Here a is that of formula (1). Therefore, the spectrum of C_φ on $\mathcal{H}^2$ is a spiral (which becomes a segment if $\Im a = 0$) joining 1 with the origin.

For our purposes, it is sufficient to know the eigenfunctions of C_φ and their corresponding eigenvalues. For each non negative number $t \geq 0$ we consider the inner function

$$e_t(z) = \exp\left[t\frac{z+1}{z-1}\right]$$

that have already appeared when dealing with the parabolic automorphism in the previous chapter.

An elementary computation, using formula (2) with $n = 1$, shows that if φ is a parabolic non-automorphism, then

$$C_\varphi e_t(z) = e^{-at}e_t(z).$$

Thus, e^{-at} is an eigenvalue corresponding to the eigenfunction $e_t(z)$ and whenever $\nu < 1/4$. We point out here that it can be shown, using Lemma 4.3, that each of these eigenvalues is simple, but we will not use this fact.

The eigenfunctions form a spanning set.

The fact that the collection of eigenfunctions $\{e_t(z) : t \geq 0\}$ of C_φ spans the Hardy space $\mathcal{H}^2$ is known to many specialists. In fact, Patrick Ahern has pointed out to us that, although not explicitly stated, it is contained in the proof of Theorem 6.1 in [AC]. Zorboska has also proved the result for the Hardy space $\mathcal{H}^2$ [Zo1, Prop. 5.8]. Using Paley-Wiener Theorem for the Hardy space of the upper half plane,

Shapiro [Sh3] has added a further proof to this result for the Hardy space. Another very simple one, suggested by Donald Sarason, can be found in [GM1]. Using a different method, we will improve the result by showing that the eigenfunctions span $\mathcal{S}_\nu$ whenever they are in $\mathcal{S}_\nu$. The proof, that will be delayed, is very simple.

LEMMA 4.7. *Suppose that $\nu < 1/4$. Then $\overline{\text{span}}\{e_t(z) : t \geq 0\} = \mathcal{S}_\nu$. In other words, the collection of functions $\{e_t(z) : t \geq 0\}$ spans $\mathcal{S}_\nu$ if and only if they are in $\mathcal{S}_\nu$.*

Renorming $\mathcal{S}_\nu$.

To obtain good orthogonality properties related to the set of functions $\{e_t : t \geq 0\}$ in the spaces $\mathcal{S}_\nu$, it is strictly necessary to modify slightly the norm on the spaces $\mathcal{S}_\nu$. For $f(z) = \sum_{k=0}^{\infty} a_k z^k \in \mathcal{S}_\nu$ we redefine the norm in $\mathcal{S}_\nu$ by

$$\|f\|_\nu^2 = |a_0|^2 + \sum_{n=1}^{\infty} |a_n|^2 n^{2\nu}. \tag{8}$$

Since the norm above is equivalent to the usual norm in $\mathcal{S}_\nu$ and hypercyclicity is invariant under similarity, it will be sufficient to prove Theorem 4.6 with this new norm. We point out that for the Hardy space the norm just defined is nothing else but the usual one. With the new inner product that induces the norm above we can prove the following lemma, which is the key point in the proof of Theorem 4.6. In all that follows, if $\mathcal{H}_i$, $i = 1, 2$, are Hilbert spaces, then $\mathcal{H}_1 \oplus \mathcal{H}_2$ will always denote orthogonal sum.

LEMMA 4.8. *Let $\{\nu_n\}_{n \geq 0}$ be defined by $\nu_n = -(2^n - 1)/2$ and let $\tau > 0$ be fixed. Suppose that $\mathcal{S}_{\nu_n}$ is endowed with the inner product that induces the norm in (8). Then, for each non negative integer n, there is an infinite dimensional subspace $Y_{\tau,n} \subset \overline{\text{span}}\{e_t : t > \tau\}$ such that*

$$\mathcal{S}_{\nu_n} = \overline{\text{span}}\{e_t : 0 \leq t \leq \tau\} \oplus Y_{\tau,n}.$$

In Chapter 5, see Theorem 5.8 and Remark 5.9, it will be shown that the statement of Lemma 4.8 is also true for any space $\mathcal{S}_\nu$ with $\nu < 1/4$.

Lemma 4.8 will be proved later on. Now, we are in position to prove Theorem 4.6. Shapiro's proof of the fact that λC_φ is not hypercyclic in $\mathcal{H}^2$ for any complex number λ is based on finding a C_φ invariant subspace in which λC_φ is similar to C_φ acting on $\mathcal{H}^2$, which is not hypercyclic. There are some differences in the proof below because we do not know that C_φ is not hypercyclic for spaces greater than the Bergman space.

PROOF OF THEOREM 4.6. By the Comparison Principle it is enough to prove that λC_φ is not hypercyclic on each of the spaces $\mathcal{S}_{\nu_k}$, where the ν_k's are given by Lemma 4.8. We proceed by a way of contradiction. Assume that λC_φ is hypercyclic on $\mathcal{S}_{\nu_k}$. Let τ be fixed. By Lemma 4.8 we can write

$$\mathcal{S}_{\nu_k} = \overline{\text{span}}\{e_t : 0 \leq t \leq \tau\} \oplus Y_{\tau,k},$$

where $Y_{\tau,k}$ is contained in $\overline{\text{span}}\{e_t : t \geq \tau\}$. Using the facts that $\overline{\text{span}}\{e_t : 0 \leq t \leq \tau\}$ is invariant under λC_φ and $Y_{\tau,k}$ is not the null subspace, it is easy to show that $\lambda C_\varphi|_{Y_{\tau,k}}$ is hypercyclic on $Y_{\tau,k}$ (see [He, Prop. 2.2]).

Let M_{e_τ} denote the operator of multiplication by e_τ defined on $\mathcal{S}_\nu$ ($\nu \leq 0$). Using the norm furnished by Lemma 1.2 ($l = 0$) and the fact that $|e_\tau(z)| < 1$, it is

easy to see that M_{e_τ} is bounded on $\mathcal{S}_\nu$. On the other hand, the following identity holds
$$e_\tau \mathcal{S}_\nu = e_\tau \overline{\text{span}}\{e_t : t \geq 0\} = \overline{\text{span}}\{e_t : t \geq \tau\}.$$
Moreover, since M_{e_τ} is clearly one to one, we find that it is an isomorphism from $\mathcal{S}_\nu$ onto $\overline{\text{span}}\{e_t : t \geq \tau\}$. Hence, as the following equalities hold for any $g \in \mathcal{S}_\nu$
$$C_\varphi^n(e_\tau(z)g(z)) = C_{\varphi_n}(e_\tau(z)g(z)) = e_\tau(\varphi_n(z))g(\varphi_n(z)) = e^{-\tau a n} e_\tau(z) C_\varphi^n g(z),$$
the spectral radius of the restriction of C_φ to the invariant subspace $\overline{\text{span}}\{e_t : t \geq \tau\}$ is equal to $Re^{-\tau \Re a}$, where R is the spectral radius of C_φ acting on $\mathcal{S}_\nu$. Since, by Lemma 4.8, the space $Y_{\tau,k}$ is contained in $\overline{\text{span}}\{e_t : t \geq \tau\}$, we conclude that the spectral radius of $\lambda C_\varphi|_{Y_{\tau,k}}$ is less than or equal to $|\lambda| R e^{-\tau \Re a}$. Hence, it is sufficient to take τ large enough to force that the spectrum of $\lambda C_\varphi|_{Y_{\tau,k}}$ does not meet the unit circle. Therefore, it follows [Ki, Thm. 2.8] that $\lambda C_\varphi|_{Y_{\tau,k}}$ cannot be hypercyclic on $Y_{\tau,k}$ for τ large enough; a contradiction. $\square$

Isometric functional Hilbert spaces.

In order to prove Lemma 4.8 we need to introduce functional Hilbert spaces that will be regarded as dense subspaces of $L^2(\mathbb{R}^+, dt/t)$. The key point is that the latter subspace will play a prominent role to transfer orthogonality properties of the functions $\{e_t : t \geq 0\}$ from $\mathcal{S}_\nu$ to $\mathcal{S}_{2\nu-1/2}$.

Recall from Chapter 1 that any of the spaces $\mathcal{S}_\nu$ is a functional Hilbert space in which $X = \mathbb{D}$. However, our interest is to transform $\mathcal{S}_\nu$ in a functional Hilbert space in which it is easier to handle the functions $e_t(z)$. Also, we will have some advantage if we ignore the the constant functions. We will work on the subspace $\mathcal{S}_\nu^0$ of those $\mathcal{S}_\nu$ functions that vanish at the origin.

For each $t \in \mathbb{R}^+$ we consider the following function which is in $\mathcal{S}_\nu^0$ for all $\nu < 1/4$
$$d_t(z) = e_t(z) - \langle e_t(z), 1 \rangle = e_t(z) - e^{-t}.$$
Lemma 4.7 implies that $\mathcal{S}_\nu^0 = \overline{\text{span}}\{d_t : t > 0\}$ whenever $\nu < 1/4$. For each $f \in \mathcal{S}_\nu^0$ we define the function
$$\hat{f}(t) = \langle f, d_t(z)\rangle_\nu \qquad (t \in \mathbb{R}^+).$$
Since $|\hat{f}(t) - \hat{f}(s)| \leq \|f\|\|d_t - d_s\|$, we find that each function $\hat{f}$ is continuous on $\mathbb{R}^+$. We set
$$\widehat{\mathcal{S}}_\nu^0 = \{\hat{f} : f \in \mathcal{S}_\nu^0\}$$
and consider the linear map Ψ_ν that assigns to each function $f \in \mathcal{S}_\nu^0$ the function $\Psi_\nu(f) = \hat{f} \in \widehat{\mathcal{S}}_\nu^0$. Since $\overline{\text{span}}\{d_t : t > 0\}$ is dense in $\mathcal{S}_\nu^0$, it follows that Ψ_ν is one-to-one from $\mathcal{S}_\nu^0$ onto $\widehat{\mathcal{S}}_\nu^0$. Furthermore, with the inner product
$$\langle \hat{f}, \hat{g}\rangle = \langle f, g\rangle,$$
the space of functions $\widehat{\mathcal{S}}_\nu^0 = \{\hat{f} = \Psi_\nu(f) : f \in \mathcal{S}_\nu^0\}$ becomes a Hilbert space and Ψ_ν is an isometric isomorphism from $\mathcal{S}_\nu^0$ onto $\widehat{\mathcal{S}}_\nu^0$. In addition, for $t \in \mathbb{R}^+$, the pointwise evaluation
$$\hat{f}(t) = \langle \hat{f}, \hat{d}_t \rangle$$
is a bounded linear functional. Thus $\widehat{\mathcal{S}}_\nu^0$ satisfies Definition 1.1 and, therefore, it is a functional Hilbert space. Observe that norm convergence in $\widehat{\mathcal{S}}_\nu^0$ implies pointwise convergence.

A complete orthogonal system of $L^2(\mathbb{R}^+, dt/t)$.

Although the work in this subsection is inspired in the previous one, it does not depend on the functional Hilbert spaces just defined. Also, it does not depend on Lemma 4.7 either. Thus the proof of Lemma 4.7 at the end of the subsection will be consistent.

We set $f_0(z) = 1$ and for $n \geq 1$ we set $f_n(z) = z^n/n^{2\nu}$. Let D_z^n the n-th partial derivative of f with respect to z. For any non negative integer n we set

$$\hat{f}_n(t) = \langle f_n(z), e_t(z)\rangle_\nu = \frac{1}{n^{2\nu}}\langle z^n, e_t(z)\rangle_\nu = \frac{1}{n!} D_z^n e_t(z)\Big|_{z=0}.$$

Observe that while the f_n's depend on ν, the $\hat{f}_n$'s do not. We write down the first of these functions

$$\hat{f}_0(t) = e^{-t}, \quad \hat{f}_1(t) = -2te^{-t}, \quad \hat{f}_2(t) = (-2t + 2t^2)e^{-t}, \ldots.$$

We have

PROPOSITION 4.9. *For each $n \geq 1$ the function $\hat{f}_n(t) = p_n(t)e^{-t}$, where p_n is a polynomial of degree n that vanishes at the origin. Moreover, the following recursion formula is satisfied*

$$\hat{f}_n(t) = -2t \sum_{k=0}^{n-1} \left(1 - \frac{k}{n}\right) \hat{f}_k(t).$$

PROOF. Upon applying Leibnitz's rule in the second equality below we obtain the recursion formula

$$\hat{f}_n = \frac{1}{n!} D_z^n(e_t(z))\Big|_{z=0}$$
$$= \frac{1}{n!} D_z^{n-1}\left(\frac{-2t}{(1-z)^2} e_t(z)\right)\Big|_{z=0}$$
$$= \frac{-2t}{n!} \sum_{k=0}^{n-1} \binom{n-1}{k} (D_z^k e_t(z) D_z^{n-1-k}[(1-z)^{-2}])\Big|_{z=0}$$
$$= -2t \sum_{k=0}^{n-1} \left(\frac{n-k}{n}\right) \frac{1}{k!} D_z^k e_t(z)\Big|_{z=0}$$
$$= -2t \sum_{k=0}^{n-1} \left(1 - \frac{k}{n}\right) \hat{f}_k.$$

Now, the rest of the statement of the proposition follows immediately. □

The following theorem is fundamental for the rest of the work in this section.

THEOREM 4.10. *The sequence of functions $\{\hat{f}_n\}_{n\geq 1}$ is a complete orthogonal system of $L^2(\mathbb{R}^+, dt/t)$ and*

$$\|\hat{f}_n\|_{L^2(\mathbb{R}^+, dt/t)}^2 = \int_0^\infty |\hat{f}_n|^2 \frac{dt}{t} = \frac{1}{n}. \tag{9}$$

PROOF. First we prove that $\{\hat{f}_n\}_{n\geq 1}$ is an orthogonal system satisfying (9). To this end, we start by showing that for $n \geq m \geq 0$ the following is true

$$\int_0^\infty \hat{f}_n(t)\hat{f}_m(t)\,dt = \begin{cases} 1/2, & \text{if } n = m = 0; \\ 1 & \text{if } n = m \neq 0; \\ -1/2, & \text{if } n - m = 1; \\ 0, & \text{otherwise.} \end{cases} \tag{10}$$

The proof of (10) will be by induction on n. Since $\hat{f}_0(t) = e^{-t}$ and $\hat{f}_1(t) = -2te^{-t}$, the above assertion for $n = 0$ and $n = 1$ follows from easy computations.

Now suppose that we have already proved (10) for $0 \leq k \leq n-1$. We use integration by parts. We set

$$u(t) = \hat{f}_n(t) = \frac{1}{n!} D_z^n e_t(z)\Big|_{z=0}.$$

Hence,

$$\begin{aligned}
u'(t) &= D_t\left(\frac{1}{n!}D_z^n(e_t(z))\Big|_{z=0}\right) \\
&= \frac{1}{n!}D_z^n(D_t e_t(z))\Big|_{z=0} \\
&= \frac{1}{n!}D_z^n\left(\frac{z+1}{z-1}e_t(z)\right)\Big|_{z=0} \\
&= \frac{1}{n!}\sum_{k=0}^n \binom{n}{k}\left(D_z^k e_t(z) D_z^{n-k}\left[\frac{z+1}{z-1}\right]\right)\Big|_{z=0} \\
&= -2\sum_{k=0}^{n-1}\frac{1}{k!}D_z^k e_t(z)\Big|_{z=0} - \frac{1}{n!}D_z^n e_t(z)\Big|_{z=0} \\
&= -2\sum_{k=0}^{n-1}\hat{f}_k - \hat{f}_n.
\end{aligned}$$

On the other hand, we set

$$v'(t) = \hat{f}_m(t) = \frac{1}{m!}D_z^m(e_t(z))\Big|_{z=0}$$

and, therefore, upon integrating with respect to t we find that

$$\begin{aligned}
v(t) &= \frac{1}{m!}D_z^m\left(\frac{z-1}{z+1}e_t(z)\right)\Big|_{z=0} \\
&= \frac{1}{m!}\sum_{k=0}^m \binom{m}{k}\left(D_z^k e_t(z) D_z^{m-k}\left[\frac{z-1}{z+1}\right]\right)\Big|_{z=0} \\
&= -2\sum_{k=0}^{m-1}(-1)^{m-k}\frac{1}{k!}D_z^k e_t(z)\Big|_{z=0} - \frac{1}{m!}D_z^m e_t(z)\Big|_{z=0} \\
&= -2\sum_{k=0}^{m-1}(-1)^{m-k}\hat{f}_k - \hat{f}_m,
\end{aligned}$$

where, if $m = 0$, the summation in the last display does not appear. Now, we use induction on m. If $m = 0$, integration by parts yields

$$\int_0^\infty \hat{f}_n(t)\hat{f}_0(t)\, dt = (uv)\Big|_0^\infty - 2\sum_{k=0}^{n-1}\int_0^\infty \hat{f}_k(t)\hat{f}_0(t)\, dt - \int_0^\infty \hat{f}_n(t)\hat{f}_0(t)\, dt.$$

The first term in the right-hand side above is zero because u and v vanish at 0 and $+\infty$. Therefore, using our induction hypothesis we have

$$\int_0^\infty \hat{f}_n\hat{f}_0\, dt = \sum_{k=0}^{n-1}\int_0^\infty \hat{f}_k\hat{f}_0\, dt = \int_0^\infty \hat{f}_0^2\, dt + \int_0^\infty \hat{f}_1\hat{f}_0\, dt = \frac{1}{2} - \frac{1}{2} = 0.$$

Now, suppose that for a fixed positive integer n we have proved that (10) is true for all k with $0 \leq k < m$. We can proceed as in the case $m = 0$ to obtain that (10) is equal to

$$-2\sum_{j=0}^{m-1}(-1)^{m-j}\sum_{i=0}^{n-1}\int_0^\infty \hat{f}_i\hat{f}_j\, dt - \sum_{i=0}^{n-1}\int_0^\infty \hat{f}_i\hat{f}_m\, dt - \sum_{j=0}^{m-1}(-1)^{m-j}\int_0^\infty \hat{f}_j\hat{f}_n\, dt. \quad (11)$$

Now, if $0 < m < n-1$ the induction hypothesis shows that each of the terms above is zero. If $m = n-1$, as before, the first and third term in (11) are zero. As for the second term we have

$$-\sum_{i=0}^{n-1}\int_0^\infty \hat{f}_i\hat{f}_{n-1}\, dt = -\int_0^\infty \hat{f}_{n-2}\hat{f}_{n-1}\, dt - \int_0^\infty \hat{f}_{n-1}^2\, dt = \frac{1}{2} - 1 = -\frac{1}{2}.$$

Finally, if $m = n$, the second and third term in (11) cancel and the first term is equal to

$$2\int_0^\infty \hat{f}_{n-2}\hat{f}_{n-1}\, dt + 2\int_0^\infty \hat{f}_{n-1}^2\, dt = -1 + 2 = 1.$$

Therefore, the double induction is complete and (10) holds.

To finish the proof, we use the recursion formula furnished by Proposition 4.9. Now, if $n > m$, using (10), we have

$$\int_0^\infty \hat{f}_n\hat{f}_m\frac{dt}{t} = -2\sum_{k=0}^{n-1}\left(1 - \frac{k}{n}\right)\int_0^\infty \hat{f}_k\hat{f}_m\, dt = 1 - \frac{m-1}{n} - 2 + \frac{2m}{n} + 1 - \frac{m+1}{n} = 0.$$

In a similar way we compute

$$\int_0^\infty \hat{f}_n^2\frac{dt}{t} = -2\sum_{k=0}^{n-1}\left(1 - \frac{k}{n}\right)\int_0^\infty \hat{f}_k\hat{f}_n = 1 - \frac{n-1}{n} = \frac{1}{n}$$

as required.

It remains to show that $\{\hat{f}_n\}_{n\geq 1}$ is a complete system. This is a consequence of the fact that $\hat{f}_n(t) = p_n(t)e^{-t}$ where p_n is a polynomial of degree n that vanishes at the origin. Now, these functions span $L^2(\mathbb{R}^+, dt/t)$ if and only if the functions

$$\{q_n(t)t^{1/2}e^{-t}, n \geq 1 \text{ and } q_n(t) = p_n(t)/t\}$$

span $L^2(\mathbb{R}^+, dt)$. But this holds because span$\{q_n : n \geq 1\}$ contains the set of all polynomials which are dense in $L^2((1/k, k), dt)$ for all positive integer k. $\square$

The extension of the isometries.
We are know in position to prove Lemma 4.7.

PROOF OF LEMMA 4.7. We will regard $\mathcal{S}_\nu$ as a linear subspace of $\mathcal{S}_{2\nu-1/2}$. Of course, this is only possible for $2\nu - 1/2 \leq \nu$, that is, $\nu \leq 1/2$. We consider the orthogonal sum $\widetilde{\mathcal{S}}_{2\nu-1/2} = [\hat{f}_0] \oplus L^2(\mathbb{R}^+, dt/t)$ where $[\hat{f}_0]$ denotes the unidimensional space generated by $\hat{f}_0$. Then, we consider the linear map

$$\widetilde{\Psi}_\nu \sum_{n=0}^\infty a_n z^n = \sum_{n=0}^\infty a_n n^{2\nu} \hat{f}_n(t).$$

Using Theorem 4.10, we have

$$\left\| \sum_{n=1}^\infty a_n n^{2\nu} \hat{f}_n \right\|_{L^2(\mathbb{R}^+, dt/t)}^2 = \sum_{n=1}^\infty |a_n|^2 n^{4\nu} \int_0^\infty \hat{f}_n^2(t) \frac{dt}{t}$$

$$= \sum_{n=1}^\infty |a_n|^2 n^{4\nu-1}$$

$$= \left\| \sum_{n=1}^\infty a_n z^n \right\|_{2\nu-1/2}^2.$$

We can conclude that $\widetilde{\Psi}_\nu$ defines an isometric isomorphism from $\mathcal{S}_{2\nu-1/2}$ onto $\widetilde{\mathcal{S}}_{2\nu-1/2}$. Furthermore, if $f(z) = \sum_{n=0}^\infty a_n z^n \in \mathcal{S}_\nu$, then

$$\langle f, e_t \rangle_\nu = \sum_{n=0}^\infty a_n n^{2\nu} \langle z^n / n^{2\nu}, e_t \rangle_\nu = \sum_{n=0}^\infty a_n n^{2\nu} \hat{f}_n = \widetilde{\Psi} f.$$

Consequently, if $f \in \mathcal{S}_\nu$ is orthogonal in $\mathcal{S}_\nu$ to e_t for all $t \geq 0$, then $\widetilde{\Psi}_\nu f$ is the zero function and, therefore, so is f. Now, the statement of the lemma follows because span $\{e_t : t \geq 0\}$ is a linear space. $\square$

Now, we can reverse the point of view of the proof of Lemma 4.7 to show that $\Psi_\nu : \mathcal{S}_0 \longrightarrow \widehat{\mathcal{S}}_\nu^0$ extends to an isometric isomorphism from $\mathcal{S}_{2\nu-1/2}^0$ onto $L^2(\mathbb{R}^+, dt/t)$. Observe that if $f_n(z) = z^n/n^{2\nu}$, then

$$\langle f_n, d_t \rangle_\nu = \langle f_n, d_t \rangle_\nu = \hat{f}_n(t) \qquad (n \geq 1)$$

is also the complete orthogonal system in Theorem 4.10.

THEOREM 4.11. *For any $\nu < 1/4$ the map $\Psi_\nu : \mathcal{S}_\nu^0 \longrightarrow \widehat{\mathcal{S}}_\nu^0$ extends to an isometric isomorphism $\widetilde{\Psi}_\nu$ from $\mathcal{S}_{2\nu-1/2}^0$ onto $L^2(\mathbb{R}^+, dt/t)$.*

PROOF. As in the proof of Lemma 4.7, for $f(z) = \sum_{n=1}^\infty a_n z^n \in \mathcal{S}_\nu^0$ we have

$$\hat{f} = \Psi_\nu f = \sum_{n=1}^\infty a_n n^{2\nu} \hat{f}_n.$$

and

$$\|\hat{f}\|_{L^2(\mathbb{R}^+, dt/t)}^2 = \|f\|_{2\nu-1/2}^2.$$

Since Ψ_ν is one to one from $\mathcal{S}_\nu^0$ onto $\widehat{\mathcal{S}}_\nu^0$ and both spaces are dense in $\mathcal{S}_{2\nu-1/2}^0$ and $L^2(\mathbb{R}^+, dt/t)$, respectively, we find that Ψ_ν extends in a unique way to an isometric isomorphism $\widetilde{\Psi}_\nu$ from $\mathcal{S}_{2\nu-1/2}^0$ onto $L^2(\mathbb{R}^+, dt/t)$. Therefore, the statement of the theorem follows. □

REMARK 4.12. If we had used the set of functions $\{e_t : t \geq 0\}$ to transform $\mathcal{S}_\nu$ onto a functional Hilbert space of functions on $\mathbb{R}^+ \cup \{0\}$ that we denote by $\widehat{\mathcal{S}}_\nu$, then we would have obtained that the corresponding isometry $\Psi_\nu : \mathcal{S}_\nu \to \widehat{\mathcal{S}}_\nu$ extends to an isometric isomorphism from $\mathcal{S}_{2\nu-1/2}$ onto $\mathbb{C} \oplus L^2(\mathbb{R}^+, dt/t)$. In this way, the proof of Lemma 4.8 below would be slightly less simple.

The Laguerre Polynomials.

The method of proof of Theorem 4.10 is, of course, known to the experts in special functions (see [Ra, Chap. 12], for instance). In fact, once we know that the set of functions $\{\hat{f}_n\}$ is a complete orthogonal system, one can ask if this set of functions corresponds to a classical orthogonal system. The answer is almost positive. In fact, if we set $\hat{f}_n(t) = p_n(t)e^{-t}$, where p_n is a polynomial of degree n, then $p_n(t) = -(2t/n)L_{n-1}^{(1)}(2t)$, where $L_n^{(1)}(t)$ is the Laguerre polynomial of degree n and of index 1 (see [Ra, Chap. 12]). We remark here that the sequence

$$\mathcal{L}_{n-1}^{(1)}(2t) = (2/\sqrt{n})L_{n-1}^{(1)}(2t)\sqrt{t}e^{-t} \qquad (n \geq 1)$$

is a complete orthonormal system of $L^2(\mathbb{R}^+, dt)$.

To prove Lemma 4.8 we need a theorem about summability of the Fourier series of certain orthogonal systems that consists of very regular functions. Roughly speaking: "the quicker the Fourier coefficients of a function f goes to zero, the more regular the function f is and viceversa" (see [Al, Chapter IV], [Sz, Chapter IX] and [Ze, Chapter 9]). Let $\mathcal{S}(\mathbb{R})$ denote the space of rapidly decreasing functions on $\mathbb{R}$, that is, the space of functions $f : \mathbb{R} \to \mathbb{C}$ that are infinitely differentiable and for all integers $m, n \geq 0$

$$\sup\{t^m |f^{n)}(t)| : t \in \mathbb{R}\} < \infty.$$

Let $\mathcal{S}^+(\mathbb{R})$ be the space of functions on $\mathbb{R}^+$ that have an extension to some function in $\mathcal{S}(\mathbb{R})$. Let ℓ denote the space of rapidly decreasing sequences, that is, the space of complex sequences $\{b_n\}$ such that for any polynomial p

$$\sum_{n=0}^{\infty} |b_n p(n)| < \infty.$$

The following theorem is due to Durán [Du, Thm. 2.5].

THEOREM 4.13. *A function f is in $t^{1/2}\mathcal{S}^+$ if and only if its Fourier coefficients are in ℓ.*

With this theorem in hand, we are in position to prove Lemma 4.8.

PROOF OF LEMMA 4.8. We will denote by $[f]$ the one dimensional subspace generated by the function f. First we will prove that there are two infinite dimensional subspaces $X_{\tau,n}$ and $Y_{\tau,n}$ and a finite dimensional subspace $D_{\tau,n}$ such

that
$$\begin{aligned}\mathcal{S}^0_{\nu_n} &= X_{\tau,n} \oplus D_{\tau,n} \oplus Y_{\tau,n} \\ &= \overline{\text{span}}\{d_t : 0 < t \leq \tau\} \oplus Y_{\tau,n} \\ &= X_{\tau,n} \oplus \overline{\text{span}}\{d_t : t \geq \tau\}.\end{aligned} \qquad (12)$$

Since $\mathcal{S}_{\nu_n} = [1] \oplus \mathcal{S}^0_{\nu_n}$ and $\overline{\text{span}}\{e_t : 0 \leq t \leq \tau\} = [1] \oplus \overline{\text{span}}\{d_t : 0 < t \leq \tau\}$, it follows that
$$\mathcal{S}_{\nu_n} = \overline{\text{span}}\{e_t : 0 \leq t \leq \tau\} \oplus Y_{\tau,n}.$$

Therefore, once we have proved (12), the statement of the lemma will follow as soon as we prove that $Y_{\tau,n}$ is contained in $\overline{\text{span}}\{e_t : t \geq \tau\}$.

The proof of the decompositions in (12) will be done by induction. Let us start by proving the result for $\mathcal{S}^0_0 = \mathcal{H}^2_0$. To this end, we compute $\langle e_s(z), e_t(z) \rangle$ in the Hardy space. Thus suppose that $0 \leq t \leq s$, we have

$$\langle e_s(z), e_t(z) \rangle = \frac{1}{2\pi}\int_{-\pi}^{\pi} e_s(e^{i\theta})\overline{e_t(e^{i\theta})}\,d\theta = \frac{1}{2\pi}\int_{-\pi}^{\pi} e_{s-t}(e^{i\theta})\,d\theta = e^{t-s}.$$

Next, for $0 < t < \tau$, by using the Gram-Schmidt method, we orthonormalize d_t and d_τ. We have
$$f_t = d_t - \frac{\langle d_t, d_\tau \rangle}{\|d_\tau\|^2} d_\tau = d_t - \frac{e^{t-\tau} - e^{-t-\tau}}{1 - e^{-2\tau}} d_\tau.$$

Analogously, we also orthogonalize d_t and d_τ for $t > \tau$. We have
$$f_t = d_t - \frac{\langle d_t, d_\tau \rangle}{\|d_\tau\|^2} d_\tau = d_t - \frac{e^{\tau-t} - e^{-t-\tau}}{1 - e^{-2\tau}} d_\tau.$$

Now, the point is that, for $0 < t < \tau$, the function f_t is orthogonal to d_s for all $s \geq \tau$. Indeed, we have

$$\begin{aligned}\langle f_t, d_s \rangle &= e^{t-s} - e^{-t-s} - \frac{(e^{t-\tau} - e^{-t-\tau})(e^{\tau-s} - e^{-\tau-s})}{1 - e^{-2\tau}} \\ &= e^{t-s} - e^{-t-s} - \frac{e^{t-s}(1 - e^{-2\tau}) - e^{-t-s}(1 - e^{-2\tau})}{1 - e^{-2\tau}} \\ &= 0\end{aligned} \qquad (13)$$

Also, for $t > \tau$ each function f_t is orthogonal to d_s for all $0 < s \leq \tau$. Indeed, we have

$$\begin{aligned}\langle f_t, d_s \rangle &= e^{s-t} - e^{-t-s} - \frac{(e^{\tau-t} - e^{-t-\tau})(e^{s-\tau} - e^{-\tau-s})}{1 - e^{-2\tau}} \\ &= e^{s-t} - e^{-t-s} - \frac{e^{s-t}(1 - e^{-2\tau}) - e^{-t-s}(1 - e^{-2\tau})}{1 - e^{-2\tau}} \\ &= 0.\end{aligned} \qquad (14)$$

The subspaces we are looking for are
$$\begin{aligned}X_{\tau,0} &= \overline{\text{span}}\{f_t : 0 < t < \tau\} \quad \text{and} \\ Y_{\tau,0} &= \overline{\text{span}}\{f_t : t > \tau\}.\end{aligned}$$

The relations in (13) and (14) imply that
$$\langle f_t, f_s \rangle = 0 \qquad \text{for } 0 < t < \tau < s.$$

Therefore, by linearity and the continuity of the inner product, we also have
$$\langle f, g \rangle = 0 \qquad (f \in X_{\tau,0},\ g \in Y_{\tau,0}).$$
As $\mathcal{H}_0^2 = \overline{\operatorname{span}}\{d_t : t > 0\}$, we have the following decomposition that summarizes all we have proved so far
$$\mathcal{H}_0^2 = X_{\tau,0} \oplus [d_\tau] \oplus Y_{\tau,0}. \tag{15}$$
From the orthogonal decomposition above, one easily checks that the following orthogonal decompositions also hold
$$\mathcal{H}_0^2 = X_{\tau,0} \oplus \overline{\operatorname{span}}\{d_t : t \geq \tau\} = \overline{\operatorname{span}}\{d_\tau : 0 < t \leq \tau\} \oplus Y_{\tau,0}. \tag{16}$$
Therefore, we have obtained all the orthogonal decompositions in (12) for $\mathcal{H}_0^2$.

For the sake of clarity we also show step 2 of the induction, that is, the decompositions in (12) hold for the Bergman space $\mathcal{S}_{-1/2}^0$. Let Ψ_0 be the isometric isomorphism between $\mathcal{S}_0^0$ and $\widehat{\mathcal{S}}_0^0$. For any set Z that is contained in $\mathcal{S}_0^0$ we write $\widehat{Z} = \Psi_0(Z)$. The orthogonal decompositions in (16) along with the definition of Ψ_0 imply that
$$\begin{cases} \text{For } \hat{f} \in \widehat{X}_{\tau,0}, & \hat{f}(s) = 0 \quad \text{for each } s \geq \tau. \\ \text{For } \hat{f} \in \widehat{Y}_{\tau,0}, & \hat{f}(s) = 0 \quad \text{for each } 0 < s \leq \tau. \end{cases} \tag{17}$$
For the sake of clarity, when a function $\hat{f} \in \widehat{\mathcal{S}}_0^0$ is considered as a function in $L^2(\mathbb{R}^+, dt/t)$ we will write $\tilde{f}$ instead of $\hat{f}$. We can consider the following subspaces of $L^2(\mathbb{R}^+, dt/t)$,
$$\widetilde{X}_{\tau,0} = \overline{\operatorname{span}}^{L^2}\{\tilde{f} : \text{such that } \hat{f} \in \widehat{X}_{\tau,0}\} \qquad \text{and}$$
$$\widetilde{Y}_{\tau,0} = \overline{\operatorname{span}}^{L^2}\{\tilde{f} : \text{such that } \hat{f} \in \widehat{Y}_{\tau,0}\}.$$
From (17) we deduce
$$\begin{cases} \text{For } \tilde{f} \in \widetilde{X}_{\tau,0}, & \tilde{f}(s) = 0 \quad \text{for almost every } s \geq \tau. \\ \text{For } \tilde{g} \in \widetilde{Y}_{\tau,0}, & \tilde{g}(s) = 0 \quad \text{for almost every } 0 < s \leq \tau. \end{cases}$$
Consequently,
$$\langle \tilde{f}, \tilde{g} \rangle_{L^2} = \int_0^\infty \tilde{f}(t)\overline{\tilde{g}(t)} \frac{dt}{t} = 0 \qquad (f \in \widetilde{X}_{\tau,0}, g \in \widetilde{Y}_{\tau,0}).$$
Since the functions in $\widehat{X}_{\tau,0}$ and $\widehat{Y}_{\tau,0}$ have disjoint supports, it follows that
$$\overline{\operatorname{span}}^{L^2}\{\tilde{f} : \text{such that } \hat{f} \in \widehat{X}_{\tau,0} \oplus \widehat{Y}_{\tau,0}\} = \widetilde{X}_{\tau,0} \oplus \widetilde{Y}_{\tau,0}.$$
Moreover, from (15) we see that $\widehat{X}_{\tau,0} \oplus \widehat{Y}_{\tau,0}$ is the orthogonal complement to $\hat{d}_\tau$ with respect to the inner product in $\widehat{\mathcal{S}}_0^0$ and, therefore, it consists of all functions in $\widehat{\mathcal{S}}_0^0$ that vanish at τ. Since the set of all polynomials that vanish at τ is dense in $L^2([\alpha, \beta], dt)$ for every pair of real numbers α, β satisfying $0 < \alpha < \tau < \beta < \infty$, we may conclude
$$L^2(\mathbb{R}^+, dt/t) = \overline{\operatorname{span}}^{L^2}\{\tilde{f} : \text{such that } \hat{f} \in \widehat{X}_{\tau,0} \oplus \widehat{Y}_{\tau,0}\}.$$
Therefore,
$$L^2(\mathbb{R}^+, dt/t) = \widetilde{X}_{\tau,0} \oplus \widetilde{Y}_{\tau,0}.$$

In particular, we must have
$$\widetilde{X}_{\tau,0} = L^2((0,\tau), dt/t) \quad \text{and} \quad \widetilde{Y}_{\tau,0} = L^2((\tau,\infty), dt/t).$$

Now, let us consider the orthogonal decomposition
$$\tilde{d}_\tau = \tilde{g}_\tau \oplus \tilde{h}_\tau \qquad (\tilde{g}_\tau \in \widetilde{X}_{\tau,0},\ \tilde{h}_\tau \in \widetilde{Y}_{\tau,0}).$$

Let $\widetilde{X}_{\tau,1}$ be the orthogonal complement of $\tilde{g}_\tau$ in $\widetilde{X}_{\tau,0} = L^2((0,\tau), dt/t)$ and let $\widetilde{Y}_{\tau,1}$ be the orthogonal complement of $\tilde{h}_\tau$ in $\widetilde{Y}_{\tau,0} = L^2((\tau,\infty), dt/t)$. We set $\widetilde{D}_{\tau,1} = [\tilde{g}_t] \oplus [\tilde{h}_t]$. Thus we have
$$L^2(\mathbb{R}^+, dt/t) = \widetilde{X}_{\tau,1} \oplus \widetilde{D}_{\tau,1} \oplus \widetilde{Y}_{\tau,1}.$$

It is easy to check that $\widetilde{X}_{\tau,1} \oplus \widetilde{D}_{\tau,1} = \overline{\operatorname{span}}\,\{\tilde{d}_t : 0 < t \leq \tau\}$ and $\widetilde{D}_{\tau,1} \oplus \widetilde{Y}_{\tau,1} = \overline{\operatorname{span}}\,\{\tilde{d}_t : t \geq \tau\}$. Therefore, we also have the orthogonal decompositions
$$L^2(\mathbb{R}^+, dt/t) = \overline{\operatorname{span}}\,\{\tilde{d}_t : 0 < t \leq \tau\} \oplus \widetilde{Y}_{\tau,1} = \widetilde{X}_{\tau,1} \oplus \overline{\operatorname{span}}\,\{\tilde{d}_t : t \geq \tau\}.$$

Finally, we consider $\widetilde{\Psi}_0^{-1} : L^2(\mathbb{R}^+, dt/t) \longrightarrow \mathcal{S}_{-1/2}^0$, where $\widetilde{\Psi}_0$ is the extension of Ψ_0 given by Theorem 4.11. Then, by just taking $X_{\tau,1} = \widetilde{\Psi}^{-1}(\widetilde{X}_{\tau,1})$, $Y_{\tau,1} = \widetilde{\Psi}^{-1}(\widetilde{Y}_{\tau,1})$ and $D_{\tau,1} = \widetilde{\Psi}^{-1}(\widetilde{D}_{\tau,1})$ we see that $\mathcal{S}_{-1/2}^0$ satisfies the orthogonal decompositions in (12).

Now, suppose that the orthogonal decompositions in (12) hold for $\mathcal{S}_{\nu_n}^0$. We will prove that the same is true for $\mathcal{S}_{\nu_{n+1}}^0$. Let Ψ_{ν_n} be the isometric isomorphism that transform $\mathcal{S}_{\nu_n}^0$ onto the functional Hilbert space $\widehat{\mathcal{S}}_{\nu_n}^0$. Our induction hypothesis shows that
$$\begin{cases} \text{If } \hat{f} \in \widehat{X}_{\tau,n}, \text{ then} & \hat{f}(s) = 0 \quad \text{for each } s \geq \tau. \\ \text{If } \hat{f} \in \widehat{Y}_{\tau,n}, \text{ then} & \hat{f}(s) = 0 \quad \text{for each } 0 < s \leq \tau. \end{cases} \tag{18}$$

Next, we consider the inclusion $i : \widehat{\mathcal{S}}_{\nu_n}^0 \to L^2(\mathbb{R}^+, dt/t)$, defined by $i(\hat{f}) = \tilde{f}$, where $\tilde{f}$ is only defined almost everywhere. We define
$$\widetilde{X}_{\tau,n} = \overline{\operatorname{span}}^{L^2}\{\tilde{f} : \text{ such that } \hat{f} \in \widehat{X}_{\tau,n}\} \quad \text{and}$$
$$\widetilde{Y}_{\tau,n} = \overline{\operatorname{span}}^{L^2}\{\tilde{f} : \text{ such that } \hat{f} \in \widehat{Y}_{\tau,n}\}.$$

Since the functions in $\widehat{X}_{\tau,n}$ and $\widehat{Y}_{\tau,n}$ are of disjoint support, it is easily seen that
$$\overline{\operatorname{span}}^{L^2}\{\tilde{f} : \hat{f} \in \widehat{X}_{\tau,n} \oplus \widehat{Y}_{\tau,n}\} = \widetilde{X}_{\tau,n} \oplus \widetilde{Y}_{\tau,n}.$$

Now, we claim that
$$L^2(\mathbb{R}^+, dt/t) = \widetilde{X}_{\tau,n} \oplus \widetilde{Y}_{\tau,n}.$$

If we show that
$$\widetilde{X}_{\tau,n} = L^2((0,\tau), dt/t) \quad \text{and} \quad \widetilde{Y}_{\tau,n} = L^2((\tau,\infty), dt/t),$$

then our claim follows immediately. From (18), we see that $\widehat{X}_{\tau,n}$ consists of all functions in $L^2(\mathbb{R}^+, dt/t)$ that vanish on $[\tau, \infty)$ and such that
$$\hat{f} = \sum_{k=1}^\infty a_k k^{2\nu_n} \hat{f}_k \quad \text{with} \quad \sum_{k=1}^\infty |a_k|^2 k^{2\nu_n} < \infty.$$

Now, recall that $\hat{f}_k(t) = -(2t/k)L_{k-1}^{(1)}(2t)\exp(-t)$ where $L_k^{(1)}(t)$ is the Laguerre polynomial of degree k and of index 1 and that the sequence of functions

$$\mathcal{L}_{k-1}^{(1)}(2t) = \frac{2}{k^{1/2}}L_{k-1}^{(1)}(2t)\sqrt{t}\exp(-t) \qquad (k \geq 1)$$

is a complete orthonormal system of $L^2(\mathbb{R}^+, dt)$. Let M_ϕ denote the multiplication operator by $\phi(t) = 1/\sqrt{t}$. As $M_\phi(f_k/\|f_k\|) = -\mathcal{L}_{k-1}(2t)$, we find that M_ϕ induces an isometric isomorphism from $L^2(\mathbb{R}^+, dt/t)$ onto $L^2(\mathbb{R}^+, dt)$. Let $\widehat{Z}$ denote the image of $\widehat{X}_{\tau,n}$ under M_ϕ. It is clear that $\widehat{Z}$ is the set of functions that vanish on $[\tau, \infty)$ and such that

$$\hat{f}(t) = \sum_{k=1}^{\infty} a_k k^{2\nu_n - 1/2}\mathcal{L}_{k-1}^{(1)}(2t) \quad \text{with} \quad \sum_{k=1}^{\infty}|a_k|^2 k^{2\nu_n} < \infty. \tag{19}$$

Let f be an infinitely differentiable function whose support is contained in $(0, \tau)$. Then f is also in $\mathcal{S}(\mathbb{R}^+)$. Thus by Theorem 4.13 its Fourier coefficients satisfies (19). Therefore, $\widehat{Z}$ contains the set of all differentiable functions whose support is contained in $(0, \tau)$. Since these latter set is dense in $L^2((0, \tau), dt)$, we may conclude that $\widetilde{X}_{\tau,n} = L^2((0, \tau), dt/t)$. The fact that $\widetilde{Y}_{\tau,n} = L^2((\tau, \infty), dt/t)$ follows in a similar way.

Now, let $\widetilde{G}_n$ and $\widetilde{H}_n$ be the orthogonal projections of $\widetilde{D}_{\tau,n}$ onto $\widetilde{X}_{\tau,n}$ and $\widetilde{Y}_{\tau,n}$, respectively. Clearly, $\widetilde{G}_n$ and $\widetilde{H}_n$ are of finite dimension. Let $\widetilde{X}_{\tau,n+1}$ be the orthogonal complement of $\widetilde{G}_n$ in $\widetilde{X}_{\tau,n}$ and let $\widetilde{Y}_{\tau,n+1}$ be the orthogonal complement of $\widetilde{H}_n$ in $\widetilde{Y}_{\tau,n}$. Hence, we have the orthogonal decomposition

$$L^2(\mathbb{R}^+, dt/t) = \widetilde{X}_{\tau,n+1} \oplus \widetilde{G}_n \oplus \widetilde{H}_n \oplus \widetilde{Y}_{\tau,n+1}.$$

We set $\widetilde{D}_{\tau,n+1} = \widetilde{G}_n \oplus \widetilde{H}_n$. It is not difficult to see that the following decompositions are also true

$$\begin{aligned} L^2(\mathbb{R}^+, dt/t) &= \widetilde{X}_{\tau,n+1} \oplus \widetilde{D}_{\tau,n+1} \oplus \widetilde{Y}_{\tau,n+1} \\ &= \overline{\text{span}}\,\{\tilde{d}_t : 0 < t \leq \tau\} \oplus \widetilde{Y}_{\tau,n+1} \\ &= \widetilde{X}_{\tau,n+1} \oplus \overline{\text{span}}\,\{\tilde{d}_t : t \geq \tau\}. \end{aligned}$$

Finally, by just considering $\widetilde{\Psi}_{\nu_{n+1}}^{-1} : L^2(\mathbb{R}^+, dt/t) \longrightarrow \mathcal{S}_{\nu_{n+1}}^0$ one can easily completes the induction.

It remains to prove that $Y_{\tau,n}$ is contained in $\overline{\text{span}}^{\mathcal{S}_{\nu_n}}\{e_t : t \geq \tau\}$. By just analyzing the proof of (12) one easily checks that $Y_{\tau,n}$ is a subspace of $\overline{\text{span}}^{\mathcal{S}_{\nu_n}}\{Y_{\tau,0}\}$. Therefore, it will be enough to show the result for the Hardy space, that is, $Y_{\tau,0}$ is a subspace of $\overline{\text{span}}^{\mathcal{H}^2}\{e_t : t \geq \tau\}$. Toward this end, we consider for each $t > \tau$ the function

$$g_t = e_t - \langle e_t, e_\tau \rangle e_\tau = e_t - e^{\tau-t}e_\tau,$$

which is orthogonal to e_τ. Moreover, each function g_t is also orthogonal to e_s for any $0 \leq s \leq \tau$. Indeed, we have

$$\langle e_s, g_t \rangle = e^{s-\tau} - e^{\tau-t}e^{s-\tau} = 0.$$

Therefore, if we set $Y_\tau = \overline{\text{span}}\{g_t : t > \tau\}$, then, using Lemma 4.7, one easily obtains the orthogonal decomposition

$$\mathcal{H}^2 = \overline{\text{span}}\{e_t : 0 \leq t \leq \tau\} \oplus Y_\tau.$$

On the other hand, we also have from (13) that
$$\mathcal{H}^2 = \overline{\operatorname{span}}\{e_t : 0 \leq t \leq \tau\} \oplus Y_{\tau,0}.$$

It follows, from the uniqueness of the orthogonal complement of a subspace, that $Y_{\tau,0} = Y_\tau$. Since Y_τ is clearly contained in $\overline{\operatorname{span}}\{e_t : t \geq \tau\}$, the desired result follows. The proof of Lemma 4.8 and that of Theorem 4.6 is now completed. $\square$

REMARK 4.14. The choice of the ν_k's in the statement of Lemma 4.7 is by no means by chance. For instance, it is not difficult to show that for $-1/2 < \nu < 0$ the functions f_t and f_s for $t < \tau < s$ are not orthogonal in $\mathcal{S}_\nu^0$. Moreover, they are not orthogonal in $\mathcal{S}_{-1/2}^0$ when using the usual norm in $\mathcal{S}_{-1/2}^0$.

REMARK 4.15. The proof of Lemma 4.7 gives a little more. It proves that $D_{\tau,n}$ is at most of dimension 2^n. Actually, $D_{\tau,n}$ has exactly dimension 2^n. Indeed, the decomposition in (12) along with the definition of $\widehat{X}_{\tau,n}$ shows that the restriction of each function in $\widehat{D}_{\tau,n}$ to (τ, ∞) is different from the null function. The corresponding situation is also true for $(0, \tau)$. Consequently, each of the functions in $\widetilde{D}_{\tau,n}$ has non zero projections onto $L^2((0,\tau), dt/t)$ as well as onto $L^2((\tau, \infty), dt/t)$.

REMARK 4.16. We stress here that it can be proved, following the same lines of [GS, Section 6], that $X = \operatorname{span}\{e_t : 0 \leq t \leq \tau\}$ is dense in the space of all holomorphic functions $\mathcal{H}(\mathbb{D})$ and the same is true for $Y = \operatorname{span}\{e_t : t \geq \tau\}$. As in [GS, Section 6], one can prove that the hypercyclicity of C_φ in the whole space of holomorphic functions is a consequence of the fact that X and Y are dense in $\mathcal{H}(\mathbb{D})$. The above remark suggests that X and Y are still dense in a weighted Hardy space $\mathcal{H}^2(\beta)$ that contains $\mathcal{S}_\nu$ for all ν. This allows us to conjecture that C_φ is hypercyclic on such a space $\mathcal{H}^2(\beta)$.

CHAPTER 5

SUPERCYCLIC LINEAR FRACTIONAL COMPOSITION OPERATORS

This chapter is devoted to characterizing the supercyclic linear fractional composition operators on weighted Dirichlet spaces. Recall from Chapter 1, that a bounded operator T on a Hilbert space $\mathcal{H}$ is said to be supercyclic if there is a vector $x \in \mathcal{H}$ such that $\{\lambda T^n x : n = 0, 1, \ldots$ and $\lambda \in \mathbb{C}\}$ is dense in $\mathcal{H}$. Further along this chapter we will prove the following theorem that completes Table I in the introduction.

THEOREM 5.1. *Let φ be a linear fractional self map of $\mathbb{D}$. Then, C_φ is supercyclic on $\mathcal{S}_\nu$ if and only if φ is a non elliptic automorphism of $\mathbb{D}$ and $\nu < 1/2$ or φ is a hyperbolic non-automorphism and $\nu \leq 1/2$.*

THE EASY PART

Since supercyclicity is a stronger property than cyclicity and a weaker property than hypercyclicity and the Comparison Principle also works for supercyclic operators, we can analyze each of the results in the preceding chapters to check that a major part of Theorem 5.1 follows immediately from our work in the previous chapters. Actually, to complete the proof of Theorem 5.1 we have only to determine whether C_φ is supercyclic or not in the following three cases:

(a) The symbol φ has an interior fixed point.
(b) The symbol φ is a hyperbolic non-automorphism.
(c) The symbol φ is a parabolic non-automorphism.

In this section we will dispense with cases (a) and (b). When φ has an interior fixed point (Chapter 2), Ansari and Bourdon [AB] proved that C_φ is not supercyclic on the Hardy space. This fact was obtained as a corollary of their result that no power bounded operator can be supercyclic. Here we present an elementary proof which works for any of the $\mathcal{H}^2(\beta)$ spaces. This result strengthens that of Proposition 2.10.

THEOREM 5.2. *Let $\mathcal{H}^2(\beta)$ be a weighted Hardy space. Let φ be an analytic self map of $\mathbb{D}$ with an interior fixed point and assume that C_φ acts boundedly on $\mathcal{H}^2(\beta)$. Then C_φ is not supercyclic.*

PROOF. Let p be the interior fixed point of φ. Suppose, by a way of contradiction, that f is a supercyclic vector for C_φ. It is clear that $f(p)$ must be different from 0, otherwise any vector in the projective C_φ-orbit of f would vanish at p, and this is not the case for all functions in $\mathcal{H}^2(\beta)$. Without loss of generality, we may assume that $f(p) = 1$. Suppose that a function $g \in \mathcal{H}^2(\beta)$ is in the projective orbit of f under C_φ. Then, there is a sequence $\{\lambda_{n_k}\}$ such that $\{\lambda_{n_k} C_{\varphi_{n_k}} f\}$ tends to

g in $\mathcal{H}^2(\beta)$ as k tends to ∞. Since norm convergence in $\mathcal{H}^2(\beta)$ implies pointwise convergence, we have

$$\begin{aligned} g(p) &= \lim_{k\to\infty} \lambda_{n_k}(C_{\varphi_{n_k}}f)(p) \\ &= \lim_{k\to\infty} \lambda_{n_k} f(\varphi_{n_k}(p)) \\ &= \lim_{k\to\infty} \lambda_{n_k} f(p) \\ &= \lim_{k\to\infty} \lambda_{n_k}. \end{aligned} \quad (1)$$

We must distinguish two cases:

Case (i). The map φ is not an elliptic automorphism. In this case, the sequence of iterates of φ tends uniformly on compact subsets of $\mathbb{D}$ to its Denjoy-Wolff point, that is, the interior fixed point p (see [Sh2, p. 79], for instance). Consequently, for each $z \in \mathbb{D}$, we have

$$g(z) = \lim_{k\to\infty} \lambda_{n_k} C_{\varphi_{n_k}} f(z) = \lim_{k\to\infty} \lambda_{n_k} f(\varphi_{n_k}(z)) = \lim_{k\to\infty} \lambda_{n_k} = g(p).$$

We conclude that only constant functions can be in the closure of the projective C_φ-orbit of f, a contradiction. Therefore, C_φ is not supercyclic.

Case (ii). The map φ is an elliptic automorphism. In this case, the argument above does not work because the iterates of φ do not converge to the fixed point. However, we can still prove that C_φ is not supercyclic. We have only to consider the case in which φ is conjugate to an irrational multiple of π, otherwise C_φ is not even cyclic [see Chap. 1, Thm. 2.3]. We take the function g to be equal to $\varphi(z) = C_\varphi z \in \mathcal{H}^2(\beta)$. Thus, from (1) we have

$$\varphi(z) = \lim_{k\to\infty} \lambda_{n_k} f(\varphi_{n_k}(z)) = \varphi(p) \lim_{k\to\infty} f(\varphi_{n_k}(z)) = p \lim_{k\to\infty} f(\varphi_{n_k}(z)).$$

By extracting a subsequence, if necessary, we may suppose that $\{\varphi_{n_k}\}$ converges uniformly on compact subsets of $\mathbb{D}$ to an elliptic automorphism ψ of $\mathbb{D}$. Thus, for each $z \in \mathbb{D}$, we have

$$\varphi(z) = pf \circ \psi(z).$$

We conclude that p must be different from zero and $f = (1/p)\varphi \circ \psi^{-1}$. It follows that all the scalar multiples of the C_φ-orbit of f are univalent functions and, Hurwitz's Theorem (see [Ah]) implies that they cannot approximate non univalent functions like $h(z) = z^2$ which is in any $\mathcal{H}^2(\beta)$. Thus f cannot be supercyclic either. The proof is finished. $\square$

Case (b) is when φ is a hyperbolic non-automorphism (Chapter 1). We have the following easy theorem.

THEOREM 5.3. *Let φ be a hyperbolic non-automorphism. Then C_φ acting on $\mathcal{S}_\nu$ is supercyclic if and only if $\nu \leq 1/2$.*

PROOF. If $\nu \leq 1/2$, then C_φ is supercyclic because there are scalar multiples of C_φ that are hypercyclic [Chap. 2, Thm 2.11].

Conversely, suppose that $\nu > 1/2$. Let η be the boundary fixed point of φ. The spectrum of C_φ is

$$\sigma(C_\varphi) = \{t : |t| \leq \varphi'(\eta)^{(2\nu-1)/2}\} \cup \{\varphi'(\eta)^j : j = 0, 1 \ldots\}.$$

(see [Hu, Cor. 12]). Since $\varphi'(\eta) < 1$, the spectrum $\sigma(C_\varphi)$ has, at least, two connected components and no circle centered at the origin can meet each of these connected components. Therefore, by a result of Herrero (see [He, Proposition 3.1]), C_φ cannot be supercyclic. The proof is concluded. $\square$

It remains to study case (c).

THE PARABOLIC NON-AUTOMORPHISM

In this final section we will prove the following theorem, which can be considered as an improvement of Theorem 4.6 and completes the proof of Theorem 5.1.

THEOREM 5.4. *Let φ be a parabolic non-automorphism that takes the unit disk into itself. Then C_φ is not supercyclic on any of the $\mathcal{S}_\nu$ spaces.*

As an application, observe that by choosing ν large enough Theorem 5.4 and the Comparison Principle show that C_φ is not supercyclic on the Bergman spaces $\mathcal{A}^p$, $1 \leq p < \infty$. This also shows that our results also apply to Banach spaces of analytic functions on the disk.

The Angle Criterion.

To prove Theorem 5.4 we need the following criterion that is the supercyclic version of Lemma 2.14.

LEMMA 5.5 (The Angle Criterion). *Let T be a bounded linear operator on a separable Hilbert space $\mathcal{H}$. Then f is supercyclic for T if and only if for any non zero vector $g \in \mathcal{H}$*

$$\sup_n \frac{|\langle T^n f, g\rangle|}{\|T^n f\|\|g\|} = 1.$$

The proof is very simple and is based on the fact that if each of the elements in the orbit of f lies outside a cone around g, then their scalar multiples cannot approximate g. This criterion was also used in [GM1] to prove that C_φ is not supercyclic for the Hardy space $\mathcal{H}^2$. The idea of the angle was firstly used in [MS] in relation to infinite dimensional closed subspaces of supercyclic vectors for weighted shifts. This latter subject arose from the work in [Mo].

With the Angle Criterion in hand, supercyclicity becomes more manageable because the scalars in the definition of supercyclic operator disappear with the quotient.

We note here that, by the Comparison Principle, for $\nu \geq 0$ the non supercyclicity of C_φ, where φ is a parabolic non-automorphism, follows from the result for the Hardy space ($\nu = 0$). The proof for $\mathcal{H}^2$ (see [GM1]) is based on very nice orthogonality properties of the eigenfunctions of C_φ that can be used to handle finite dimensional C_φ-invariant subspaces. This along with Gershchgorin's Theorem allowed us to estimate the size of C_φ on certain infinite dimensional invariant subspaces.

Unfortunately, the orthogonality properties of the eigenfunctions of C_φ get worse as ν decreases and the techniques in [GM1] do not apply to the general situation. A suitable modification of the proof in [GM1] has allowed the authors (unpublished) to prove the corresponding statement of Theorem 5.4 for the Bergman space, $\nu = -1/2$. The proof also uses the orthogonality properties of some combinations of the eigenfunctions. However, the calculations become quite intractable for

$\nu < -1/2$. The techniques of the proof of Theorem 5.4, that we will present here, are completely different from those of the Hardy and Bergman spaces and depend, in an essential way, on the representation of the adjoint $C_\varphi^\star$ as a multiplication operator acting on a functional Hilbert space.

The adjoint of C_φ.

To prove Theorem 5.4, as for other cyclic properties (see Lemma 2.15), we can ignore the constant functions. Thus, we will work with the space $\mathcal{S}_\nu^0$ of those $\mathcal{S}_\nu$ functions that vanish at the origin. Since supercyclicity is also preserved under similarities, as in Chapter 4, we will consider $\mathcal{S}_\nu^0$ endowed with the modified norm

$$\|f\|_\nu^2 = \sum_{n=1}^\infty |a_n|^2 n^{2\nu} \quad \text{for} \quad f(z) = \sum_{n=1}^\infty a_n z^n \in \mathcal{S}_\nu^0.$$

We also may suppose that the boundary fixed point of φ is 1 and, therefore, we may assume that it is given by the formula

$$\varphi(z) = \frac{(2-a)z + a}{-az + 2 + a} \quad \text{where} \quad \Re a > 0. \tag{2}$$

On the other hand, for each $t > 0$ the eigenfunction

$$d_t = e_t(z) - e_t(0) = \exp\left[t\frac{1+z}{1-z}\right] - e^{-t}$$

corresponds to the eigenvalue e^{-at}. Finally, we recall from the previous chapter that the map

$$\Psi_\nu : \mathcal{S}_\nu^0 \to \widehat{\mathcal{S}}_\nu^0$$

defined by $\Psi_\nu(f) = \hat{f}$ is an isometry from $\mathcal{S}_\nu^0$ onto $\widehat{\mathcal{S}}_\nu^0 = \{\hat{f}(t) = \langle f, d_t \rangle : f \in \mathcal{S}_\nu^0 \text{ and } t > 0\}$. All this together allows us to prove that the adjoint $C_\varphi^\star$ is similar to a multiplication operator defined on the functional Hilbert space $\widehat{\mathcal{S}}_\nu^0$. This is, in fact, the key point for all the work in this section.

LEMMA 5.6. *Let φ be a parabolic non-automorphism that takes the unit disk into itself and suppose that it satisfies formula (2). Suppose also that $\nu < 1/4$. Then the adjoint $C_\varphi^\star$ is similar to the multiplication operator $M_\phi : \widehat{\mathcal{S}}_\nu^0 \to \widehat{\mathcal{S}}_\nu^0$, where $\phi(t) = e^{-\bar{a}t}$.*

PROOF. Let $\widehat{C}_\varphi : \widehat{\mathcal{S}}_\nu^0 \to \widehat{\mathcal{S}}_\nu^0$ be the image of C_φ under Ψ_ν, that is, $\widehat{C}_\varphi \hat{f} = \widehat{C_\varphi f}$; then

$$\begin{aligned}(\widehat{C}_\varphi^\star \hat{f})(t) &= \widehat{C_\varphi^\star f}(t) \\ &= \langle C_\varphi^\star f, d_t \rangle \\ &= \langle f, C_\varphi d_t \rangle \\ &= \langle f, e^{-at} d_t \rangle \\ &= e^{-\bar{a}t} \langle f, d_t \rangle \\ &= e^{-\bar{a}t} \hat{f}(t)\end{aligned}$$

and the statement of the lemma follows because for $\nu < 1/4$ the functions d_t span $\mathcal{S}_\nu^0$. □

REMARK 5.7. Actually, Lemma 5.6 is the realization of a general result about bounded linear operators on a Hilbert space $\mathcal{H}$, namely, an operator T has a spanning set of eigenvectors if and only if its adjoint $T^\star$ is similar to a multiplication operator on a functional Hilbert space (see [Ha, Problem 69]).

In order to apply the Angle Criterion, as in [GM1], we have to look for two subspaces in which we can control the size of the norm of the operator: in one of them C_φ must be "small" and in the other "big". Basically, from the proof of Theorem 4.6, we know that if we take $\tau > 0$ large enough, then C_φ acting on the following C_φ-invariant subspace

$$\overline{\operatorname{span}}\{d_t : t \geq \tau\}$$

has a small spectral radius. Thus, our purpose is to search for a subspace where C_φ is bounded from below, which is, actually, the main difficulty of the proof of Theorem 5.4.

In what follows, it will be convenient to consider $\widehat{\mathcal{S}}_\nu^0$ as a linear subspace of $L^2(\mathbb{R}^+, dt/t)$. By Theorem 4.10 the functions

$$\hat{f}_n(t) = \frac{1}{n!} D^n e_t(z) \qquad (n \geq 1)$$

form a complete orthogonal system of $L^2(\mathbb{R}^+, dt/t)$ and we can regard this space as the space of functions $\hat{f}(t) = \sum_{n=1}^\infty a_n \hat{f}_n(t)$ for which the norm

$$\|\hat{f}\|^2 = \sum_{n=1}^\infty \frac{|a_n|^2}{n}$$

is finite. On the other hand, the space $\widehat{\mathcal{S}}_\nu^0$ is the space of functions $\hat{f}(t) = \sum_{n=1}^\infty a_n \hat{f}_n(t)$ for which

$$\|\hat{f}\|_\nu^2 = \sum_{n=1}^\infty \frac{|a_n|^2}{n^{2\nu}}$$

is finite. It is convenient to note that, as ν gets smaller, the space $\widehat{\mathcal{S}}_\nu^0$ gets smaller while the space $\mathcal{S}_\nu^0$ gets bigger. Thus the isometries Ψ_ν's ($\nu < 1/4$) interchange the sense of the inclusions between the spaces.

It is not difficult to prove that the operator of multiplication by $e^{\bar{a}t}$, where $\Re a > 0$, is unbounded on any of the $\widehat{\mathcal{S}}_\nu^0$ spaces. However, if we truncate the space, then it becomes bounded. This is the content of the following lemma, which is a major step in the proof of Theorem 5.4. The point is that, as Halmos [Ha] observes, it is difficult to determine the boundedness of multipliers on functional Hilbert spaces.

LEMMA 5.8. *Consider the sequence $\nu_i = 1/2 - i$, $i = 1, 2, \ldots$. Let τ be any positive real number. Let $\widehat{X}_{\tau,\nu_i}$ be the subspace of those functions in $\widehat{\mathcal{S}}_{\nu_i}^0$ that vanish on the whole interval (τ, ∞). Consider the function $\phi(t) = e^{\bar{a}t}$, where a is any complex number. Then the multiplication operator M_ϕ is bounded on $\widehat{X}_{\tau,\nu_i}$.*

To prove Lemma 5.8, we need to develop some more properties of the orthogonal system $\{\hat{f}_n\}$.

A three term recurrence relation.
We begin by recalling that

$$e_t(z) = \sum_{n=0}^{\infty} \hat{f}_n(t) z^n.$$

Upon taking derivatives with respect to z we have

$$\frac{-2t}{(z-1)^2} \sum_{n=0}^{\infty} \hat{f}_n(t) z^n = \sum_{n=1}^{\infty} n \hat{f}_n(t) z^{n-1}.$$

Multiplying by $(z-1)^2$ and grouping like powers of z we obtain

$$-2t \sum_{n=0}^{\infty} \hat{f}_n(t) z^n = \sum_{n=0}^{\infty} ((n+1)\hat{f}_{n+1}(t) - 2n\hat{f}_n(t) + (n-1)\hat{f}_{n-1}(t)) z^n,$$

where it is understood that $\hat{f}_{-1}$ is the null function. Thus $\{\hat{f}_n\}$ satisfy the following three term recurrence relation

$$(n+1)\hat{f}_{n+1}(t) = (2n - 2t)\hat{f}_{n-1}(t) + (1-n)\hat{f}_{n-1}(t) \qquad (n \geq 0). \qquad (3)$$

This is not surprising because $\hat{f}_n = p_n(t) e^{-t}$ and $p_n(t) = -(2t/n) L_{n-1}^{(1)}(2t)$ where $L_{n-1}^{(1)}$ is the Laguerre polynomial of degree $n-1$ and of index 1 (see [Ra, p. 202, formula (4)]). For our purposes it is preferable to write formula (3) in the form

$$t\hat{f}_n(t) = -\frac{n+1}{2} \hat{f}_{n+1}(t) + n \hat{f}_n(t) - \frac{n-1}{2} \hat{f}_{n-1}(t) \qquad (n \geq 0). \qquad (4)$$

The usefulness of (4) will be soon apparent.

An integration formula.
We will also make use of the following formula.

$$\int_t^{\infty} e^{-x} \hat{f}_n(x) \frac{dx}{x} = \frac{1}{2t} \left(\hat{f}_n(t) - \frac{n-1}{n} \hat{f}_{n-1}(t) \right) e^{-t} \qquad (t > 0 \text{ and } n \geq 1). \qquad (5)$$

To establish formula (5) we use the beginning of the proof of Theorem 4.10. We integrate twice by parts (in the first equality below $u = \hat{f}_n$ and $v' = e^{-x}$ and in the second equality $u = \hat{f}_{n-1}$ and $v' = e^{-x}$). We have

$$\int_t^{\infty} e^{-x} \hat{f}_n(x) \, dx = -\frac{1}{2} \hat{f}_n(x) e^{-x} \Big|_t^{\infty} - \sum_{k=0}^{n-1} \int_t^{\infty} \hat{f}_k e^{-x} \, dx$$

$$= \frac{1}{2} e^{-t} \hat{f}_n(t) + \frac{1}{2} e^{-x} \hat{f}_{n-1}(x) \Big|_t^{\infty}$$

$$= \frac{1}{2} e^{-t} (\hat{f}_n(t) - \hat{f}_{n-1}(t)).$$

Therefore, using the formula furnished by Proposition 4.9 in the first and the third equalities below, we have

$$\int_t^\infty \hat{f}_n(x) \frac{dx}{x} = -2 \sum_{k=0}^{n-1} \left(1 - \frac{k}{n}\right) \int_t^\infty \hat{f}_k(x) e^{-x} \, dx$$

$$= -\sum_{k=0}^{n-1} \left(1 - \frac{k}{n}\right) (\hat{f}_k - \hat{f}_{k-1}(t)) e^{-t}$$

$$= \frac{1}{2t} \left(\hat{f}_n(t) - \frac{n-1}{n} \hat{f}_{n-1}(t)\right) e^{-t},$$

which is the desired formula.

PROOF OF LEMMA 5.8. Clearly, $M_\phi = \exp(\bar{a} M_\psi)$, where M_ψ is the operator of multiplication by $\psi(t) = t$. Thus it is sufficient to prove that M_ψ is bounded on $\widehat{X}_{\tau,\nu_i}$.

We set $\nu_0 = 1/2$. Let $\widehat{X}_{\tau,\nu_0}$ be the subspace of those functions $L^2(\mathbb{R}^+, dt/t)$ that vanish on (τ, ∞) for almost everywhere. We have

$$\cdots \subset \widehat{X}_{\tau,\nu_i} \subset \widehat{X}_{\tau,\nu_{i-1}} \subset \cdots \subset \widehat{X}_{\tau,\nu_1} \subset \widehat{X}_{\tau,\nu_0}.$$

Observe that while $\widehat{X}_{\tau,\nu_i}$, $i \geq 1$, is a functional Hilbert space, $\widehat{X}_{\tau,\nu_0}$ is not.

We will prove that M_ψ is bounded on $\widehat{X}_{\tau,\nu_i}$, $i \geq 0$, by induction. Of course, if $\hat{f}(t)$ vanishes on (τ, ∞), so does $t\hat{f}(t)$. Thus we only have to prove that the norm for $\hat{f} \in \widehat{X}_{\tau,\nu_i}$ is preserved under M_ψ. For ν_0 the situation is extremely simple and the boundedness of M_ψ is a trivial consequence of the integral representation of the norm. Indeed, if $\hat{f} \subset \widehat{X}_{\tau,\nu_0}$, then

$$\|t\hat{f}(t)\|^2_{L^2(\mathbb{R}^+, dt/t)} = \int_0^\infty |t\hat{f}(t)|^2 \frac{dt}{t}$$

$$= \int_0^\tau |t\hat{f}(t)|^2 \frac{dt}{t}$$

$$\leq \tau^2 \int_0^\infty |\hat{f}(t)|^2 \frac{dt}{t}$$

$$= \tau^2 \|\hat{f}(t)\|^2_{L^2(\mathbb{R}^+, dt/t)}.$$

Suppose now that $\hat{f}(t) = \sum_{n=1}^\infty a_n \hat{f}_n(t)$ is in $L^2(\mathbb{R}^+, dt/t)$ or $\widehat{S}^0_{\nu_i}$, $i = 1, 2, \ldots$, then, using the three term recurrence relation in (4), we can get the matrix representation of M_ψ. Indeed,

$$t\hat{f}(t) = \sum_{n=1}^\infty a_n t \hat{f}_n(t)$$

$$= \sum_{n=1}^\infty a_n \left(-\frac{n+1}{2} \hat{f}_{n+1}(t) + n\hat{f}_n(t) - \frac{n-1}{2} \hat{f}_{n-1}(t)\right)$$

$$= \sum_{n=1}^\infty n \left(a_n - \frac{1}{2} a_{n+1} - \frac{1}{2} a_{n-1}\right) \hat{f}_n(t)$$

where $a_0 = 0$. The key point that makes the proof work is that the matrix representation of M_ψ with respect to $\{\hat{f}_n\}$ is independent of the underlying space. Observe that $M_\psi = DT$ where D is an unbounded diagonal operator and T is a bounded Toeplitz operator which is normal. Suppose that we have already proved that M_ψ acts boundedly on $\widehat{X}_{\tau,\nu_{i-1}}$. This means that there is a constant C such that

$$\sum_{n=1}^\infty \frac{n^2}{n^{2\nu_{i-1}}} \left| a_n - \frac{1}{2}a_{n+1} - \frac{1}{2}a_{n-1} \right|^2 \leq C \sum_{n=1}^\infty \frac{|a_n|^2}{n^{2\nu_{i-1}}} \qquad \text{for } \hat{f} \in \widehat{X}_{\tau,\nu_{i-1}}. \qquad (6)$$

Now, for each function $\hat{f}(t) = \sum_{n=1}^\infty a_n \hat{f}_n(t) \in \widehat{X}_{\tau,\nu_i}$, we consider the function

$$\hat{g}(t) = \sum_{n=1}^\infty (n+1) a_n \hat{f}_{n+1}(t) = \sum_{n=2}^\infty n a_{n-1} \hat{f}_n(t).$$

In the computations that follow we use repeatedly that $\nu_i = \nu_{i-1} - 1$. Observe that

$$\|\hat{g}\|^2_{\nu_{i-1}} = \sum_{n=2}^\infty \frac{n^2}{n^{2\nu_{i-1}}} |a_{n-1}|^2 = \sum_{n=1}^\infty \frac{|a_n|^2}{(n+1)^{2\nu_i}} \leq 2^{-2\nu_i} \|\hat{f}\|^2_{\nu_i}. \qquad (7)$$

Thus $\hat{g}(t) \in \widehat{\mathcal{S}}^0_{\nu_{i-1}}$. Suppose for the moment that we have also proved that $\hat{g}(t)$ is in $\widehat{X}_{\tau,\nu_{i-1}}$. In such a case, we can prove the boundedness of M_ψ acting in $\widehat{X}_{\tau,\nu_i}$ as follows. Since

$$t\hat{g}(t) = \sum_{n=1}^\infty n \left(n a_{n-1} - \frac{(n+1)}{2} a_n - \frac{(n-1)}{2} a_{n-2} \right) \hat{f}_n(t),$$

the induction hypothesis along with our assumption on $\hat{g}$ allows to use (6) to show that

$$\|t\hat{g}\|^2_{\nu_{i-1}} = \sum_{n=1}^\infty \frac{1}{n^{2\nu_i}} \left| n a_{n-1} - \frac{n+1}{2} a_n - \frac{n-1}{2} a_{n-2} \right|^2$$

$$\leq C \sum_{n=2}^\infty \frac{n^2}{n^{2\nu_{i-1}}} |a_{n-1}|^2$$

$$\leq C 2^{-2\nu_i} \|\hat{f}\|^2_{\nu_i},$$

where the last inequality is due to (7). We also consider the auxiliary function

$$\hat{h}(t) = \sum_{n=1}^\infty (a_{n+1} - a_n) \hat{f}_n(t),$$

which evidently satisfies $\|\hat{h}(t)\|_{\nu_i} \leq 2\|\hat{f}(t)\|_{\nu_i}$ for all $i \geq 1$. Then, upon applying the triangle inequality in the first inequality below

$$\|t\hat{f}\|_{\nu_i} = \left(\sum_{n=1}^{\infty} \frac{1}{n^{2\nu_i}} \left|na_n - \frac{n}{2}a_{n+1} - \frac{n}{2}a_{n-1}\right|^2\right)^{1/2}$$

$$\leq \left(\sum_{n=1}^{\infty} \frac{1}{n^{2\nu_i}} \left|(n+1)a_n - \frac{n+2}{2}a_{n+1} - \frac{n}{2}a_{n-1}\right|^2\right)^{1/2} + \|\hat{h}\|_{\nu_i}$$

$$= \left(\sum_{n=2}^{\infty} \frac{1}{(n-1)^{2\nu_i}} \left|na_{n-1} - \frac{n+1}{2}a_n - \frac{n-1}{2}a_{n-2}\right|^2\right)^{1/2} + \|\hat{h}\|_{\nu_i}$$

$$\leq \|t\hat{g}\|_{\nu_{i-1}} + \|\hat{h}\|_{\nu_i}$$

$$\leq C^{1/2} 2^{-\nu_i} \|\hat{f}\|_{\nu_i} + 2\|\hat{f}\|_{\nu_i}$$

$$= (C^{1/2} 2^{-\nu_i} + 2)\|\hat{f}\|_{\nu_i}.$$

Therefore, if $\hat{g} \in \widehat{X}_{\tau,\nu_{i-1}}$, then M_ψ is bounded on $\widehat{X}_{\tau,\nu_i}$. Thus the proof will be concluded once we have proved that $\hat{g} \in \widehat{X}_{\tau,\nu_{i-1}}$.

As $\hat{g} \in \widehat{\mathcal{S}}_{\nu_{i-1}}$, it remains only to prove that $\hat{g}(t)$ is the null function on (τ, ∞). Now, $\hat{g}(t)$ vanishes on (τ, ∞) if and only if

$$G(t) = \int_t^{\infty} e^{-x} \hat{g}(x) \frac{dx}{x} = 0 \qquad \text{for all } t > \tau.$$

Using formula (5) we see that the last integral is equal to

$$\int_t^{\infty} \sum_{n=1}^{\infty} (n+1)a_n e^{-x} \hat{f}_{n+1}(x) \frac{dx}{x} = \sum_{n=1}^{\infty} (n+1)a_n \int_t^{\infty} e^{-x} \hat{f}_{n+1}(x) dx$$

$$= \sum_{n=1}^{\infty} a_n \frac{1}{2t} \left((n+1)\hat{f}_{n+1}(t) - n\hat{f}_n(t)\right) e^{-t}.$$

Again, $G(t)$ is the null function on (τ, ∞) if and only if

$$H(t) = \int_t^{\infty} G(x) dx = 0 \qquad \text{for all } t > \tau.$$

Upon applying formula (5) again we obtain

$$H(t) = \frac{1}{2} \sum_{n=1}^{\infty} a_n \int_t^{\infty} \left((n+1)\hat{f}_{n+1}(x) - n\hat{f}_n(x)\right) e^{-x} \frac{dx}{x}$$

$$= \frac{1}{2} \sum_{n=1}^{\infty} a_n \frac{1}{t} \left(\frac{n+1}{2} \hat{f}_{n+1}(t) - n\hat{f}_n(t) + \frac{n-1}{2} \hat{f}_{n-1}(t)\right) e^{-t}$$

$$= \frac{1}{2} e^{-t} \sum_{n=1}^{\infty} a_n \hat{f}_n(t)$$

$$= \frac{1}{2} e^{-t} \hat{f}(t).$$

Since $\hat{f}(t)$ vanishes on (τ, ∞) the same is true for $H(t)$ and $G(t)$ and, therefore, $\hat{g}(t)$ vanishes on (τ, ∞), as required. The proof is now finished. $\square$

PROOF OF THEOREM 5.4. First of all, by Lemma 5.6, it is enough to prove the result for the image of C_φ under Ψ_ν, that is, $\widehat{C}_\varphi : \widehat{\mathcal{S}}_\nu^0 \to \widehat{\mathcal{S}}_\nu^0$. In the proof we may suppose that ν is one of the ν_i's furnished by Lemma 5.8, since the general case follows by the Comparison Principle.

We start with some preliminaries which avoid the use of Lemma 4.8. Let τ be any positive real number. Let $\widehat{X}_\tau$ and $\widehat{Y}_\tau$ be the spaces of functions in $\widehat{\mathcal{S}}_\nu^0$ that vanish on (τ, ∞) and $(0, \tau)$, respectively. Since for $\hat{f} \in \widehat{\mathcal{S}}_\nu^0$ we have $\hat{f}(t) = \langle \hat{f}, \hat{d}_t \rangle$, the definitions of $\widehat{X}_\tau$ and $\widehat{Y}_\tau$ give the orthogonal decompositions

$$\widehat{\mathcal{S}}_\nu^0 = \widehat{X}_\tau \oplus \overline{\text{span}}\{\hat{d}_t : t > \tau\} \quad \text{and} \quad \widehat{\mathcal{S}}_\nu^0 = \overline{\text{span}}\{\hat{d}_t : 0 < t < \tau\} \oplus \widehat{Y}_\tau. \qquad (8)$$

On the other hand, we claim that

$$\widehat{\mathcal{S}}_\nu^0 = \overline{\text{span}}\{\hat{d}_t : 0 < t < \tau\} + \overline{\text{span}}\{\hat{d}_t : t > \tau\}. \qquad (9)$$

Indeed, if $\hat{f} \in \widehat{\mathcal{S}}_\nu^0$ is orthogonal to the right hand side above, then $\hat{f}(t) = 0$ for all $t > 0$, except perhaps for $t = \tau$. Since $\hat{f}$ is continuous, $\hat{f}(\tau) = 0$ and, therefore, $\hat{f}$ is the null function. Thus our claim follows. Now, from (8) and (9) one may deduce immediately that

$$\widehat{X}_\tau \subset \overline{\text{span}}\{\hat{d}_t : 0 < t < \tau\} \quad \text{and} \quad \widehat{Y}_\tau \subset \overline{\text{span}}\{\hat{d}_t : t > \tau\}.$$

As a consequence, we deduce from (8) the orthogonal relationship

$$\langle \hat{f}, \hat{g} \rangle = 0, \qquad (\hat{f} \in \widehat{X}_\tau \text{ and } \hat{g} \in \widehat{Y}_\tau). \qquad (10)$$

Now, we proceed to define the spaces we need. Consider $\tau_2 > \tau_1 > 0$ and the corresponding subspaces $\widehat{X}_{\tau_i}$ and $\widehat{Y}_{\tau_i}$, $i = 1, 2$. Since $\widehat{X}_{\tau_1}$ is contained in $\widehat{X}_{\tau_2}$, it follows that $\widehat{X}_{\tau_1}$ and $\widehat{Y}_{\tau_2}$ are orthogonal. We are interested in the orthogonal sum $\widehat{X}_{\tau_1} \oplus \widehat{Y}_{\tau_2}$. Let $\widehat{Z}_{\tau_1, \tau_2}$ be the orthogonal complement of $\widehat{X}_{\tau_1} \oplus \widehat{Y}_{\tau_2}$. From (8) we see that

$$\widehat{Z}_{\tau_1, \tau_2} = \overline{\text{span}}\{d_t : t > \tau_1\} \cap \overline{\text{span}}\{d_t : 0 < t < \tau_2\}.$$

Since $\widehat{Z}_{\tau_1, \tau_2}$ is intersection of $\widehat{C}_\varphi$-invariant subspaces, it follows that it is itself $\widehat{C}_\varphi$-invariant.

Let P denote the orthogonal projection onto $\widehat{X}_{\tau_1} \oplus \widehat{Y}_{\tau_2}$ with kernel $\widehat{Z}_{\tau_1, \tau_2}$. Recall that the compression of $\widehat{C}_\varphi$ to $\widehat{X}_{\tau_1} \oplus \widehat{Y}_{\tau_2}$ is defined as $\widetilde{C}_\varphi = P\widehat{C}_\varphi = P\widehat{C}_\varphi P$. The latter equality is due to the fact that $\ker P$ is $\widehat{C}_\varphi$-invariant.

Now, we will show that $\widehat{X}_{\tau_1}$ and $\widehat{Y}_{\tau_2}$ are reducing subspaces of $\widetilde{C}_\varphi$, that is, they are invariant under $\widetilde{C}_\varphi$ as well as under $\widetilde{C}_\varphi^\star$. To prove this, we begin by computing the adjoint $\widetilde{C}_\varphi^\star$. Lemma 5.6 asserts that $\widehat{C}_\varphi^\star$ is the multiplication operator M_ϕ, where $\phi(t) = e^{-\bar{a}t}$. This is still true for $\widetilde{C}_\varphi^\star$. Indeed, due to the fact that $\widehat{X}_{\tau_1} \oplus \widehat{Y}_{\tau_2}$ is M_ϕ-invariant, we have

$$(\widetilde{C}_\varphi^\star \hat{f})(t) = (PM_\phi P\hat{f})(t) = (PM_\phi \hat{f})(t) = (M_\phi \hat{f})(t) = e^{-\bar{a}t}\hat{f}(t) \quad \text{for } \hat{f} \in \widehat{X}_{\tau_1} \oplus \widehat{Y}_{\tau_2}.$$

Thus $\widehat{X}_{\tau_1}$ and $\widehat{Y}_{\tau_2}$ are invariant under $\widetilde{C}_\varphi^\star$. In addition, since $\widehat{X}_{\tau_1}$ is orthogonal to $\widehat{Y}_{\tau_2}$, we have

$$\langle \widetilde{C}_\varphi f, g \rangle = \langle f, \widetilde{C}_\varphi^\star g \rangle = 0 \qquad (\hat{f} \in \widehat{X}_{\tau_1} \text{ and } \hat{g} \in \widehat{Y}_{\tau_2}).$$

Thus $\widehat{X}_{\tau_1}$ is also $\widetilde{C}_\varphi$-invariant and, analogously, we can see that the same is true for $\widehat{Y}_{\tau_2}$. Therefore, both subspaces are reducing subspaces.

Now, we estimate the norm of $\widetilde{C}_\varphi$ on $\widehat{X}_{\tau_1}$ and $\widehat{Y}_{\tau_2}$. First, as in the proof of Theorem 4.6, we may obtain

$$\|\widetilde{C}_\varphi \hat{f}\| \leq e^{-\tau_2 \Re a} R \qquad \text{for } \hat{f} \in \widehat{Y}_{\tau_2}, \tag{11}$$

where R is the spectral radius of $\widehat{C}_\varphi$ on $\widehat{S}^0_\nu$. On the other hand, by Lemma 5.3, the multiplication operator $\widetilde{C}^\star_\varphi$ is invertible on $\widehat{X}_{\tau_1}$ and, therefore, the same is true for $\widetilde{C}_\varphi$. Thus $\widetilde{C}_\varphi$ is bounded below on $\widehat{X}_{\tau_1}$. Hence, there is a positive constant C such that

$$\|\widetilde{C}_\varphi \hat{f}\| \geq C\|\hat{f}\| \qquad \text{for } f \in \widehat{X}_{\tau_1}. \tag{12}$$

Finally, suppose that $\widehat{C}_\varphi$ is supercyclic. Then, since $\widehat{Z}_{\tau_1,\tau_2}$ is $\widehat{C}_\varphi$-invariant, it follows, by a result of Herrero [He2], that $\widetilde{C}_\varphi$ is also a supercyclic operator. Thus, let us suppose that $\hat{f} = \hat{f}_{\tau_1} \oplus \hat{f}_{\tau_2} \in \widehat{X}_{\tau_1} \oplus \widehat{Y}_{\tau_2}$ is a supercyclic vector for $\widetilde{C}_\varphi$. In particular, since $\widehat{X}_{\tau_1}$ and $\widehat{Y}_{\tau_2}$ are reducing subspaces $\hat{f}_{\tau_i}$, $i = 1, 2$ must be different from zero, otherwise $\hat{f}$ cannot be supercyclic. Let $\hat{g} \in \widehat{X}_{\tau_1}$. Using (11) and (12) in the third inequality below we have

$$\frac{|\langle \widetilde{C}_\varphi^n \hat{f}, \hat{g}\rangle|}{\|\widetilde{C}_\varphi^n \hat{f}\|\|\hat{g}\|} = \frac{|\langle \widetilde{C}_\varphi^n \hat{f}_{\tau_1}, \hat{g}\rangle|}{\|\widetilde{C}_\varphi^n \hat{f}_{\tau_1} \oplus \widetilde{C}_\varphi^n \hat{f}_{\tau_2}\|\|\hat{g}\|}$$

$$\leq \frac{\|\widetilde{C}_\varphi^n \hat{f}_{\tau_1}\|}{\|\widetilde{C}_\varphi^n \hat{f}_{\tau_2}\|}$$

$$\leq \frac{R^n e^{-n\Re a \tau_2}}{C^n},$$

which goes to zero if τ_2 is large enough. Therefore, upon applying the Angle Criterion we see that $\hat{f}$ is not supercyclic for $\widetilde{C}_\varphi$; a contradiction. The proof of Theorem 5.4 is now finished.

With this result, our characterization of the different cyclic properties of linear fractional composition operators on weighted Dirichlet spaces is complete.

REMARK 5.9. At the beginning of the proof of Theorem 5.4 we have seen a very simple proof of almost all the statement of Lemma 4.8 and for any ν and not just for the sequence $\nu_n = (2^n - 1)/2$. The problem is that we have not proved that $\widehat{X}_\tau$ and $\widehat{Y}_\tau$ are of infinite dimension. But this is a consequence, for instance, of the fact that infinitely differentiable functions with appropriate support are in $\widehat{X}_\tau$ and $\widehat{Y}_\tau$. We maintained the original proof of Lemma 4.8 not only because it is appealing, but also because it is of interest in its own right. In particular, as remarked at the end of Chapter 4, it gives exactly the dimension of $D_{\tau,n}$, the orthogonal complement to $X_{\tau,n} \oplus Y_{\tau,n}$.

Endnotes

As was already pointed out in the preface, this work can be regarded as a first approach to a more comprehensive analysis of the different cyclic properties of general composition operators, that is, composition operators induced by any holomorphic self map of the unit disk $\mathbb{D}$. Here we outline in some detail the Linear Fractional Model Theorem that hopefully will be a fundamental tool to solve the general case.

The Denjoy-Wolff Theorem asserts that every holomorphic self map φ of $\mathbb{D}$, except the elliptic automorphisms, has an attractive fixed point to which the sequence of iterates $\{\varphi_n\}$ converges: if there is not a fixed point in $\mathbb{D}$, then there is a unique boundary fixed point that does the job. This is called the Denjoy-Wolff point (See [CM, Chap. 2] or [Sh2, Chap. 5]). Again not only the location of the Denjoy-Wolff point but also the derivative φ' near it should play a prominent role in all the analysis. Particular instances have already appeared in this work (Theorem 5.2).

The Denjoy-Wolff point provides a classification of the holomorphic self maps of $\mathbb{D}$ into four types: maps with interior fixed point, hyperbolic maps, parabolic automorphism maps and parabolic non automorphism maps. Of course, each linear fractional self map of $\mathbb{D}$ falls into one of these classes just determined by the properties of its fixed points.

With the classification in the paragraph above, it is interesting to determine the relationship between the cyclic behavior of composition operators induced by general holomorphic self maps of $\mathbb{D}$ and linear fractional composition operators on weighted Dirichlet spaces or even on weighted Hardy spaces. In particular, whether or not they share the same cyclic behavior. We already remarked in the preface that for the Hardy space, Bourdon and Shapiro made an extensive study of such a relation (see [BS2]). In this connection, we point out that the supercyclicity of composition operators induced by holomorphic (non linear fractional) self maps of $\mathbb{D}$ has not been studied even in the Hardy space.

In addition, we can direct a study of the cyclic behavior to a particular space. For instance, this work suggests that one of the spaces where an interesting study could arise is the Dirichlet space. On the opposite side, this work also suggests that a general study can be done on weighted Hardy spaces. For instance, the statements of Theorems 2.1, 2.2, 5.2, etc. lack of any hypothesis on the sequence of weights; a rare fact in the context of weighted Hardy spaces.

The Linear Fractional Model Theorem, a result of classical function theory, allows us to classify holomorphic self-maps of the unit disc in one of the classes already mentioned. This striking theorem asserts that, under very general conditions, an analytic map φ that takes $\mathbb{D}$ into itself can be intertwined with a linear fractional transformation (see [CM, Theorem 2.53]). Since the cyclic behavior of

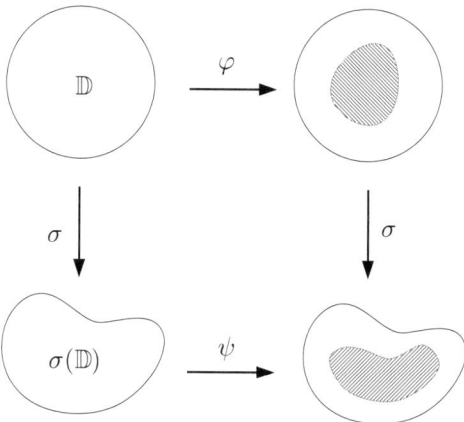

FIGURE 3

composition operators has to do only with univalent inducing symbols, then a more restricted version can be stated: If φ is a univalent self map of $\mathbb{D}$, then there exist a univalent map σ on $\mathbb{D}$ and a linear fractional self-map Φ of $\mathbb{D}$, having the same type as φ, such that
$$\sigma \circ \varphi = \Phi \circ \sigma. \qquad (1)$$
It follows from the equation above that Φ takes the simply connected domain $G = \sigma(\mathbb{D})$ into itself. Thus σ establishes a conjugacy between φ acting on $\mathbb{D}$ and the linear fractional map Φ acting on G. This interwining relation is represented in Figure 3. The pair (Φ, G) is called a linear fractional model for φ.

The importance of the Linear Fractional Model Theorem arises from the fact that many problems involving iteration or solution of functional equations can be explicitly solved for linear fractional transformations and the intertwining relates these answers to solutions of the problems for a given function φ.

Statement and solution of particular instances of the Linear Fractional Theorem trace back more than a century ago to the germinal work of Schroeder [Scr]. In 1884, Koenigs [Koe] solved the dilation case, in which equation (1) is called the Schroeder equation
$$\sigma \circ \varphi = \lambda \sigma,$$
where $\lambda = \varphi'(0)$. The solution σ was obtained as a concrete limit and it is called Koegnigs' function. This result along with the idea of translating the problem to the upper half plane allowed Valiron to solve the hyperbolic case (see [Va]). Later, in 1979 Baker and Pomerenke [BaP], and independently Cowen [Cw1] in 1981, solved the parabolic cases. Finally, using techniques that involve Riemann surfaces constructions, Cowen [Cw1] proved the Linear Fractional Model Theorem for all the cases at once. Thus, in some sense, this theorem seems to be the path to follow for determining the extent to which composition operators share the same cyclic behavior on the weighted Dirichlet spaces as their linear fractional models.

Bibliography

[AC] P. R. Ahern and D. N. Clark, *On functions orthogonal to invariant subspaces*, Acta Math. **124** (1970), 191–204.

[Ah] L. V. Ahlfors, *Complex Analysis*, McGraw-Hill, New York, 1979.

[Al] G. Alexits, *Convergence problems of orthogonal series*, Pergamon Press. New York, 1961.

[AB] S. I. Ansari and P. S. Bourdon, *Some properties of cyclic operators*, Acta Sci. Math. (Szeged) **63** (1997), 195–207.

[BaP] I.N. Baker and Ch. Pomerenke, *On the iteration of the analytic functions in a half-plane II*, J. London Math. Soc. (2) **20** (1979), 255–258..

[BP] J. Bes and A. Peris, Hereditarily Hypercyclic Operators and the Hypercyclicity Criterion, J. Funct. Anal. **167** (1999), 94–112.

[Bi] G. D. Birkhoff, *Démonstration d'un théorème élémentaire sur les fonctions entières*, C. R. Acad. Sci. Paris **189** (1929), 473-475.

[Bo] P. S. Bourdon, *Invariant manifolds of hypercyclic vectors*, Proc. Amer. Math. Soc. **118** (1993), 845-847.

[BS1] P. S. Bourdon and J. H. Shapiro, *Cyclic composition operators on H^2*, Proc. Symp. Pure Math. **51** Part 2 (1990), 43-53.

[BS2] P. S. Bourdon and J. H. Shapiro, *Cyclic phenomena for composition operators*, Mem. Amer. Math. Soc. **596** (1997).

[CS] K. C. Chan and J. H. Shapiro, *The cyclic behavior of translation operators on Hilbert spaces of entire functions*, Indiana Univ. Math. J. **40** (1991), 1421-1449.

[CR] K. F. Clancey and D. D. Rogers, *Cyclic vectors and seminormal operators*, Indiana Univ. Math. J. **27** (1978), 689–696.

[Co] J. B. Conway, *A Course in Functional Analysis*, Springer-Verlag, New York, 1985.

[Cw1] C. C. Cowen, *Iteration and solution of functional equations for functions analytic in the unit disc*, Trans. Amer. Math. Soc. **256** (1981), 69–95.

[Cw2] C. C. Cowen, *Composition operators on $\mathcal{H}^2$*, J. Operator Theory **9** (1983), 77–106.

[Cw3] C. C. Cowen, *Linear fractional composition operators on $\mathcal{H}^2$*, Integral Equations and Operator Theory **11** (1988), 151–160.

[CM] C. Cowen and B. MacCluer, *Composition Operators on Spaces of Analytic Functions*, CRC Press, 1995.

[Du] A. J. Durán, *Laguerre expansions of tempered distributions and generalized functions*, J. Math. Anal. Appl. **150** (1990), 166–180.

[Dur] P. L. Duren, *Theory of H^p spaces*, Academic Press, New York, 1970.

[GM1] E. A. Gallardo-Gutiérrez and A. Montes-Rodríguez, *The role of the Angle in the supercyclic behavior*, J. Funct. Anal. (to appear)..

[GM2] E. A. Gallardo-Gutiérrez and A. Montes-Rodríguez, *Cyclic linear fractional composition operators*, Extracta Mathematicae **15** (2000), 1–6.

[GLM] M. González, F. León and A. Montes-Rodríguez, *Semi-Fredholm Theory: hypercyclic and supercyclic subspaces*, Proc. London Math. Soc. (2000), 169–189.

[GS] R. M. Gethner and J. H. Shapiro, *Universal vectors for operators on spaces of holomorphic functions*, Proc. Amer. Math. Soc. **100** no.2 (1987), 281-288.

[Gr] K.G Große-Erdmann, *Holomorphic Monster and universelle Funktionen*, Mitt. Math. Sem. Giessen **176** (1987), 1-84.

[Ha] P. R. Halmos, *A Hilbert space problem book*, Van Nostrand Company, 1967.

[He1] D. A. Herrero, *Approximation of Hilbert space operators, Vol. I (Second edition), Pitman Research Notes in Mathematics Series 224*, Longman Scientific and Technical, Harlow, copublished in U. S. A. with John Willey and Sons, Inc., New York, 1989.

[He2] D. A. Herrero, *Limits of hypercyclic and supercyclic operators*, J. Funct. Anal. **99** (1991), 179–190.

[HK] D. A. Herrero and C. Kitai, *Invertible hypercyclic operators*, Proc. Amer. Math. Soc. **116** (1992), 873–875.

[Hi] W. M. Higdon, *Spectra of composition operators on the Dirichlet space*, preprint.

[HW] H. M. Hilden and L. J. Wallen, *Some cyclic and non-cyclic vectors of certain operators*, Indiana Univ. Math. J. **23** (1974), 557–565.

[Hu] P. R. Hurst, *Relating composition operators on different weighted Hardy spaces*, Arch. Math. **68** (1997), 503–513.

[Ki] C. Kitai, *Invariant Closed Sets for Linear Operators*, Thesis, Univ. Toronto, 1982.

[Koe] G. Koegnigs, *Recherches sur les intégrales de certaines équationes functionelles*, Annale Ecole Normale Superior (Supplément) **3** (1884), 3–41.

[LM1] F. León-Saavedra and A. Montes-Rodríguez, *Linear structure of hypercyclic vectors*, J. Funct. Anal. **148** (1997), 524-545.

[LM2] F. León-Saavedra and A. Montes-Rodríguez, *Spectral theory and hypercyclic subspaces*, Trans. Amer. Math. Soc. **353** (2001), 247–267.

[Mo] A. Montes-Rodríguez, *Banach spaces of hypercyclic vectors*, Michigan Math. J. **43** (1996), 419–436.

[MS] A. Montes-Rodríguez and H. N. Salas, *Supercyclic subspaces: spectral theory and weighted shifts*, Adv. Math. **163** (2001), 74–134.

[NS] D. J. Newman and H. S. Shapiro, *Embedding theorems for weighted classes of harmonic and analytic functions*, Michigan Math. J. **9** (1962), 249–243.

[No] E. A. Nordgren, *Composition operators*, Canadian J. Math. **20** (1968), 442–449.

[NRR] E. A. Nordgren, H. Radjavi and P. Rosenthal, *Composition operators and the invariant subspace problem*, C. R. Mat. Rep. Acad. Sci. Canada **6** (1984), 279–282.

[NRW] E. A. Nordgren, P. Rosenthal and F. S. Wintrobe, *Invertible composition operators on H^p*, J. Funct. Anal. **73** (1987), 324–344.

[RR] H. Radjavi and P. Rosenthal, *Invariant subspaces*, Springer-Verlag New York, 1973.

[Ra] E. D. Rainville, *Special functions*, Chelsea Publ. Company, New York, 1971.

[Ro] S. Rolewicz, *On orbits of elements*, Studia Math. **32** (1969), 17-22.

[Ru1] W. Rudin, *Real and complex analysis*, third ed., McGraw Hill, New York, 1987.

[Ru2] W. Rudin, *Fourier analysis on groups*, John Wiley & Sons, Inc., 1962.

[Sa] H. N. Salas, *Supercyclicity and weighted shifts*, Studia Math. **135** (1999), 55–74.

[Scr] E. Schroeder, *Über itierte Funktionen*, Math. Ann. **3** (1871), 296–322.

[SW] W. P. Seidel and J. L. Walsh, *On Approximation by Euclidean and non-Euclidean Translates of an Analytic Function*, Bull. Amer. Math. Soc. **47** (1941), 916–920.

[SS] H. S. Shapiro and A. L. Shields, *On the zeros of functions with finite Dirichlet integral and some related function spaces*, Math. Zeitcschr. **80** (1962), 217–229.

[Sh1] J. H Shapiro, *The essential norm of a composition operator*, Annals Math. **125** (1987), 375-404.

[Sh2] J. H. Shapiro, *Composition Operators and Classical Function Theory*, Springer-Verlag, 1993.

[Sh3] J. H. Shapiro, *No hypercyclicity of λC_φ with φ parabolic non-automorphism.*, Seminar Michigan State University: Hypercyclic and subnormal operators, 1998.

[Sh4] J. H. Shapiro, *Decomposability and the cyclic behavior of parabolic composition operators*, Preprint (2000).

[Shi] A. L. Shields, *Weighted shift operators and analytic function theory*, Topics in Operator Theory, Math. Surveys Monographs **13** (1974), Amer. Math. Soc., Providence, RI, 49-128.

[Sz] G. Szegö, *Orthogonal Polynomials*, Amer. Math. Soc., Coloquium Publications, New York, 1959.

[Va] G. Valiron, *Sur l'iteration des fonctions holomorphes dans un demiplan*, Bull des Sci. Math. (2) **55** (1931), 105-128.

[Ze] A. H. Zemanian, *Generalized Integral Transformation*, Interscience, New York, 1968.

[Zo1] N. Zorboska, *Composition operators on weighted Hardy spaces*, Thesis, Univ. Toronto, 1988.

[Zo2] N. Zorboska, *Composition operators on S_a spaces*, Indiana Univ. Math. J. **39** (1990), 847–857.

[Zo3]   N. Zorboska, *Cyclic composition operators on smooth weighted Hardy spaces*, Rocky Mountain J. Math. **29** (1999), 725–740.

Editorial Information

To be published in the *Memoirs*, a paper must be correct, new, nontrivial, and significant. Further, it must be well written and of interest to a substantial number of mathematicians. Piecemeal results, such as an inconclusive step toward an unproved major theorem or a minor variation on a known result, are in general not acceptable for publication. Papers appearing in *Memoirs* are generally longer than those appearing in *Transactions*, which shares the same editorial committee.

As of October 1, 2003, the backlog for this journal was approximately 5 volumes. This estimate is the result of dividing the number of manuscripts for this journal in the Providence office that have not yet gone to the printer on the above date by the average number of monographs per volume over the previous twelve months, reduced by the number of volumes published in four months (the time necessary for preparing a volume for the printer). (There are 6 volumes per year, each containing at least 4 numbers.)

A Consent to Publish and Copyright Agreement is required before a paper will be published in the *Memoirs*. After a paper is accepted for publication, the Providence office will send a Consent to Publish and Copyright Agreement to all authors of the paper. By submitting a paper to the *Memoirs*, authors certify that the results have not been submitted to nor are they under consideration for publication by another journal, conference proceedings, or similar publication.

Information for Authors

Memoirs are printed from camera copy fully prepared by the author. This means that the finished book will look exactly like the copy submitted.

The paper must contain a *descriptive title* and an *abstract* that summarizes the article in language suitable for workers in the general field (algebra, analysis, etc.). The *descriptive title* should be short, but informative; useless or vague phrases such as "some remarks about" or "concerning" should be avoided. The *abstract* should be at least one complete sentence, and at most 300 words. Included with the footnotes to the paper should be the 2000 *Mathematics Subject Classification* representing the primary and secondary subjects of the article. The classifications are accessible from www.ams.org/msc/. The list of classifications is also available in print starting with the 1999 annual index of *Mathematical Reviews*. The Mathematics Subject Classification footnote may be followed by a list of *key words and phrases* describing the subject matter of the article and taken from it. Journal abbreviations used in bibliographies are listed in the latest *Mathematical Reviews* annual index. The series abbreviations are also accessible from www.ams.org/publications/. To help in preparing and verifying references, the AMS offers MR Lookup, a Reference Tool for Linking, at www.ams.org/mrlookup/. When the manuscript is submitted, authors should supply the editor with electronic addresses if available. These will be printed after the postal address at the end of the article.

Electronically prepared manuscripts. The AMS encourages electronically prepared manuscripts, with a strong preference for $\mathcal{A}_{\mathcal{M}}\mathcal{S}$-LATEX. To this end, the Society has prepared $\mathcal{A}_{\mathcal{M}}\mathcal{S}$-LATEX author packages for each AMS publication. Author packages include instructions for preparing electronic manuscripts, the *AMS Author Handbook*, samples, and a style file that generates the particular design specifications of that publication series. Though $\mathcal{A}_{\mathcal{M}}\mathcal{S}$-LATEX is the highly preferred format of TEX, author packages are also available in $\mathcal{A}_{\mathcal{M}}\mathcal{S}$-TEX.

Authors may retrieve an author package from e-MATH starting from `www.ams.org/tex/` or via FTP to `ftp.ams.org` (login as `anonymous`, enter username as password, and type `cd pub/author-info`). The *AMS Author Handbook* and the *Instruction Manual* are available in PDF format following the author packages link from `www.ams.org/tex/`. The author package can be obtained free of charge by sending email to `pub@ams.org` (Internet) or from the Publication Division, American Mathematical Society, 201 Charles St., Providence, RI 02904, USA. When requesting an author package, please specify $\mathcal{AMS}$-LaTeX or $\mathcal{AMS}$-TeX, Macintosh or IBM (3.5) format, and the publication in which your paper will appear. Please be sure to include your complete mailing address.

Sending electronic files. After acceptance, the source file(s) should be sent to the Providence office (this includes any TeX source file, any graphics files, and the DVI or PostScript file).

Before sending the source file, be sure you have proofread your paper carefully. The files you send must be the EXACT files used to generate the proof copy that was accepted for publication. For all publications, authors are required to send a printed copy of their paper, which exactly matches the copy approved for publication, along with any graphics that will appear in the paper.

TeX files may be submitted by email, FTP, or on diskette. The DVI file(s) and PostScript files should be submitted only by FTP or on diskette unless they are encoded properly to submit through email. (DVI files are binary and PostScript files tend to be very large.)

Electronically prepared manuscripts can be sent via email to `pub-submit@ams.org` (Internet). The subject line of the message should include the publication code to identify it as a Memoir. TeX source files, DVI files, and PostScript files can be transferred over the Internet by FTP to the Internet node `e-math.ams.org` (130.44.1.100).

Electronic graphics. Comprehensive instructions on preparing graphics are available at `www.ams.org/jourhtml/graphics.html`. A few of the major requirements are given here.

Submit files for graphics as EPS (Encapsulated PostScript) files. This includes graphics originated via a graphics application as well as scanned photographs or other computer-generated images. If this is not possible, TIFF files are acceptable as long as they can be opened in Adobe Photoshop or Illustrator. No matter what method was used to produce the graphic, it is necessary to provide a paper copy to the AMS.

Authors using graphics packages for the creation of electronic art should also avoid the use of any lines thinner than 0.5 points in width. Many graphics packages allow the user to specify a "hairline" for a very thin line. Hairlines often look acceptable when proofed on a typical laser printer. However, when produced on a high-resolution laser imagesetter, hairlines become nearly invisible and will be lost entirely in the final printing process.

Screens should be set to values between 15% and 85%. Screens which fall outside of this range are too light or too dark to print correctly. Variations of screens within a graphic should be no less than 10%.

Inquiries. Any inquiries concerning a paper that has been accepted for publication should be sent directly to the Electronic Prepress Department, American Mathematical Society, 201 Charles St., Providence, RI 02904, USA.

Editors

This journal is designed particularly for long research papers, normally at least 80 pages in length, and groups of cognate papers in pure and applied mathematics. Papers intended for publication in the *Memoirs* should be addressed to one of the following editors. In principle the Memoirs welcomes electronic submissions, and some of the editors, those whose names appear below with an asterisk (*), have indicated that they prefer them. However, editors reserve the right to request hard copies after papers have been submitted electronically. Authors are advised to make preliminary email inquiries to editors about whether they are likely to be able to handle submissions in a particular electronic form.

*Algebra to ROBERT GURALNICK, Department of Mathematics, University of Southern California, Los Angeles, CA 90089-1113; email: guralnic@math.usc.edu

Algebraic geometry to DAN ABRAMOVICH, Department of Mathematics, Boston University, 111 Cummington St., Boston, MA 02215; email: abramovic@bu.edu

*Algebraic number theory to V. KUMAR MURTY, Department of Mathematics, University of Toronto, 100 St. George Street, Toronto, ON M5S 1A1, Canada; email: murty@math.toronto.edu

Algebraic topology and cohomology of groups to STEWART PRIDDY, Department of Mathematics, Northwestern University, 2033 Sheridan Road, Evanston, IL 60208-2730; email: priddy@math.nwu.edu

Combinatorics and Lie theory to SERGEY FOMIN, Department of Mathematics, University of Michigan, Ann Arbor, Michigan 48109-1109; email: fomin@umich.edu

Complex analysis and complex geometry to DUONG H. PHONG, Department of Mathematics, Columbia University, 2990 Broadway, New York, NY 10027-0029; email: phong@math.columbia.edu

*Differential geometry and global analysis to LISA C. JEFFREY, Department of Mathematics, University of Toronto, 100 St. George St., Toronto, ON Canada M5S 3G3; email: jeffrey@math.toronto.edu

Dynamical systems and ergodic theory to ROBERT F. WILLIAMS, Department of Mathematics, University of Texas, Austin, Texas 78712-1082; email: bob@math.utexas.edu

*Functional analysis and operator algebras to MARIUS DADARLAT, Department of Mathematics, Purdue University, 150 N. University St., West Lafayette, IN 47907-2067; email: mdd@math.purdue.edu

*Geometric analysis to TOBIAS COLDING, Courant Institute, New York University, 251 Mercer St., New York, NY 10012; email: colding@cims.nyu.edu

*Geometric analysis to MLADEN BESTVINA, Department of Mathematics, University of Utah, 155 South 1400 East, JWB 233, Salt Lake City, Utah 84112-0090; email: bestvina@math.utah.edu

Harmonic analysis to ALEXANDER NAGEL, Department of Mathematics, University of Wisconsin, 480 Lincoln Drive, Madison, WI 53706-1313; email: nagel@math.wisc.edu

Harmonic analysis, representation theory, and Lie theory to ROBERT J. STANTON, Department of Mathematics, The Ohio State University, 231 West 18th Avenue, Columbus, OH 43210-1174; email: stanton@math.ohio-state.edu

*Logic to STEFFEN LEMPP, Department of Mathematics, University of Wisconsin, 480 Lincoln Drive, Madison, Wisconsin 53706-1388; email: lempp@math.wisc.edu

Number theory to HAROLD G. DIAMOND, Department of Mathematics, University of Illinois, 1409 W. Green St., Urbana, IL 61801-2917; email: diamond@math.uiuc.edu

*Ordinary differential equations, and applied mathematics to PETER W. BATES, Department of Mathematics, Michigan State University, East Lansing, MI 48824-1027; email: peter@math.msu.edu

*Partial differential equations to PATRICIA E. BAUMAN, Department of Mathematics, Purdue University, West Lafayette, IN 47907-1395; email: bauman@math.purdue.edu

*Probability and statistics to KRZYSZTOF BURDZY, Department of Mathematics, University of Washington, Box 354350, Seattle, Washington 98195-4350; email: burdzy@math.washington.edu

*Real analysis and partial differential equations to DANIEL TATARU, Department of Mathematics, University of California, Berkeley, Berkeley, CA 94720; email: tataru@ math.berkeley.edu

All other communications to the editors should be addressed to the Managing Editor, WILLIAM BECKNER, Department of Mathematics, University of Texas, Austin, TX 78712-1082; email: beckner@math.utexas.edu.

Titles in This Series

795 **Adam Nyman,** Points on quantum projectivizations, 2004

794 **Kevin K. Ferland and L. Gaunce Lewis, Jr.,** The $RO(G)$-graded equivariant ordinary homology of G-cell complexes with even-dimensional cells for $G = \mathbb{Z}/p$, 2004

793 **Jindřich Zapletal,** Descriptive set theory and definable forcing, 2004

792 **Inmaculada Baldomá and Ernest Fontich,** Exponentially small splitting of invariant manifolds of parabolic points, 2004

791 **Eva A. Gallardo-Gutiérrez and Alfonso Montes-Rodríguez,** The role of the spectrum in the cyclic behavior of composition operators, 2004

790 **Thierry Lévy,** Yang-Mills measure on compact surfaces, 2003

789 **Helge Glöckner,** Positive definite functions on infinite-dimensional convex cones, 2003

788 **Robert Denk, Matthias Hieber, and Jan Prüss,** $\mathcal{R}$-boundedness, Fourier multipliers and problems of elliptic and parabolic type, 2003

787 **Michael Cwikel, Per G. Nilsson, and Gideon Schechtman,** Interpolation of weighted Banach lattices/A characterization of relatively decomposable Banach lattices, 2003

786 **Arnd Scheel,** Radially symmetric patterns of reaction-diffusion systems, 2003

785 **R. R. Bruner and J. P. C. Greenlees,** The connective K-theory of finite groups, 2003

784 **Desmond Sheiham,** Invariants of boundary link cobordism, 2003

783 **Ethan Akin, Mike Hurley, and Judy A. Kennedy,** Dynamics of topologically generic homeomorphisms, 2003

782 **Masaaki Furusawa and Joseph A. Shalika,** On central critical values of the degree four L-functions for GSp(4): The Fundamental Lemma, 2003

781 **Marcin Bownik,** Anisotropic Hardy spaces and wavelets, 2003

780 **S. Marmi and D. Sauzin,** Quasianalytic monogenic solutions of a cohomological equation, 2003

779 **Hansjörg Geiges,** h-principles and flexibility in geometry, 2003

778 **David B. Massey,** Numerical control over complex analytic singularities, 2003

777 **Robert Lauter,** Pseudodifferential analysis on conformally compact spaces, 2003

776 **U. Haagerup, H. P. Rosenthal, and F. A. Sukochev,** Banach embedding properties of non-commutative L^p-spaces, 2003

775 **P. Lochak, J.-P. Marco, and D. Sauzin,** On the splitting of invariant manifolds in multidimensional near-integrable Hamiltonian systems, 2003

774 **Kai A. Behrend,** Derived ℓ-adic categories for algebraic stacks, 2003

773 **Robert M. Guralnick, Peter Müller, and Jan Saxl,** The rational function analogue of a question of Schur and exceptionality of permutation representations, 2003

772 **Katrina Barron,** The moduli space of $N = 1$ superspheres with tubes and the sewing operation, 2003

771 **Shigenori Matsumoto,** Affine flows on 3-manifolds, 2003

770 **W. N. Everitt and L. Markus,** Elliptic partial differential operators and symplectic algebra, 2003

769 **Jie Wu,** Homotopy theory of the suspensions of the projective plane, 2003

768 **R. Höpfner and E. Löcherbach,** Limit theorems for null recurrent Markov processes, 2003

767 **Po Hu,** S-modules in the category of schemes, 2003

766 **Su Gao and Alexander S. Kechris,** On the classification of Polish metric spaces up to isometry, 2003

765 **Robert Bieri and Ross Geoghegan,** Connectivity properties of group actions on non-positively curved spaces, 2003

764 **J. Spandaw,** Noether-Lefschetz problems for degeneracy loci, 2003

TITLES IN THIS SERIES

763 **Yasuyuki Kachi and Eiichi Sato,** Segre's reflexivity and an inductive characterization os hyperquadrics, 2002

762 **Leiba Rodman, Ilya M. Spitkovsky, and Hugo Woerdeman,** Abstract band method via factorization, positive and band extensions of multivariable almost periodic matrix functions, and spectral estimation, 2002

761 **Oliver Druet and Emmanuel Hebey,** The AB program in geometric analysis : Sharp Sobolev inequalities and related problems, 2002

760 **Markus Banagl,** Extending intersection homology type invarients to non-Witt spaces, 2002

759 **Donald M. Davis,** From representation theory to homotopy groups, 2002

758 **Alan Forrest, John Hunton, and Johannes Kellendonk,** Topological invariants for projection method patterns, 2002

757 **Douglas Bowman,** q-difference operators, orthogonal polynomials, and symmetric expansions, 2002

756 **José Ignacio Cogolludo-Agustín,** Topological invariants of the complement to arrangements of rational plane curves, 2002

755 **M. A. Mandell and J. P. May,** Equivariant orthogonal spectra and S-modules, 2002

754 **Edward L. Green, Idun Reiten, and Øyvind Solberg,** Dualities on generalized Koszul algebras, 2002

753 **Daniel Panazzolo,** Desingularization of nilpotent singularities in families of planar vector fields, 2002

752 **Linus Kramer,** Homogeneous spaces, Tits buildings, and isoparametric hypersurfaces, 2002

751 **Bruce Allison, Georgia Benkart, and Yun Gao,** Lie algebras graded by the root systems BC_r, $r \geq 2$, 2002

750 **Masaki Izumi and Hideki Kosaki,** Kac algebras arising from composition of subfactors: General theory and classification, 2002

749 **Nanhua Xi,** The based ring of two-sided cells of affine Weyl groups of type $\tilde{A}_{n-1}$, 2002

748 **Jürgen Ritter and Alfred Weiss,** The lifted root number conjecture and Iwasawa theory, 2002

747 **Armand Borel, Robert Friedman, and John W. Morgan,** Almost commuting elements in compact Lie groups, 2002

746 **Peter Niemann,** Some generalized Kac-Moody algebras with known root multiplicities, 2002

745 **Mikhail A. Lifshits and Werner Linde,** Approximation and entropy numbers of Volterra operators with application to Brownian motion, 2002

744 **Roger Chalkley,** Basic global relative invariants for homogeneous linear differential equations, 2002

743 **Heng Sun,** Spectral decomposition of a covering of $GL(r)$: the Borel case, 2002

742 **J. E. Gilbert, Y. S. Han, J. A. Hogan, J. D. Lakey, D. Weiland, and G. Weiss,** Smooth molecular functions and singular integral operators, 2002

741 **Francisco Santos,** Triangulations of oriented matroids, 2002

740 **Rick Durrett,** Mutual invadability implies coexistence in spatial models, 2002

739 **Georgios K. Alexopoulos,** Sub-Laplacians with drift on Lie groups of polynomial volume growth, 2002

For a complete list of titles in this series, visit the
AMS Bookstore at **www.ams.org/bookstore/**.